# SAT

## Math 2

Andy Gaus
Kathleen Morrison
XAMonline, Inc.

**Copyright © 2016**

All rights reserved. No part of the material protected by this copyright notice may be reproduced or utilized in any form or by any means, electronic or mechanical, including photocopying or recording or by any information storage and retrievable system, without written permission from the copyright holder.

To obtain permission(s) to use the material from this work for any purpose including workshops or seminars, please submit a written request to:

XAMonline, Inc.
21 Orient Avenue
Melrose, MA 02176
Toll Free: 1-800-509-4128
Email: info@xamonline.com
Web: www.xamonline.com
Fax: 1-617-583-5552

Library of Congress Cataloging-in-Publication Data
Morrison, Kathleen

SAT Math 2 / Morrison, Kathleen
   ISBN: 978-1-60787-572-7

1. SAT         2. Study Guides        3. Mathematics

**Disclaimer:**

The opinions expressed in this publication are the sole works of XAMonline and were created independently from The College Board, or other testing affiliates. Between the time of publication and printing, specific test standards as well as testing formats and website information may change that are not included in part or in whole within this product. XAMonline develops sample test questions, and they reflect similar content as on real tests; however, they are not former tests. XAMonline assembles content that aligns with test standards but makes no claims nor guarantees candidates a passing score.

Cover photo provided by iStock.com/Andrey Prokhorov; iStock.com/Photka; ©Can Stock Photo Inc./stryjek/csp27436042; ©Can Stock Photo Inc./Eraxion/csp2309923; ©Can Stock Photo Inc./kamphi/csp3144939; ©Can Stock Photo Inc./kasto/csp14163272

**Printed in the United States of America**
SAT Math 2
ISBN: 978-1-60787-572-7

# Table of Contents

**Section I: Overview of the SAT Math Level 2 Test** ................................ 5

    **Overview** ................................................................ 7
        What Are The SAT Subject Area Tests? .................................. 7
        Choosing Between Math Level 1 and Math Level 2 ....................... 7
        Areas Tested In SAT Math Level 2 ....................................... 8
        Why Take The SAT Math Level 2 Test? ................................... 8
        How Is The SAT Math Level 2 Test Scored? .............................. 9
        Should You Guess Answers on the Test? ................................ 9
        How Is The SAT Math Level 2 Exam Administered? ....................... 10
        Accommodations for Students with Disabilities ......................... 10
        Getting Ready for Test Day ............................................ 11
        Testing Tips .......................................................... 12

**Section II: Content Review for SAT Math Level 2** ............................... 13

    **Chapter 1: Number and Operations** ..................................... 15
        1.1   Order of operations ................................................ 15
        1.2   Properties of operations ........................................... 15
        1.3   Ratios ............................................................. 17
        1.4   Proportions ........................................................ 17
        1.5   Natural (counting) numbers ........................................ 19
        1.6   Properties of odd and even numbers ............................... 19
        1.7   Composite and prime numbers ..................................... 20
        1.8   Divisibility tests for natural numbers ............................... 20
        1.9   Real numbers ..................................................... 22
        1.10  Matrices .......................................................... 23
        1.11  Vectors ........................................................... 25
        1.12  Sequences and series ............................................. 26
        1.13  Imaginary numbers ............................................... 29
        1.14  Complex numbers ................................................ 29

    **Chapter 2: Algebra and Functions** ...................................... 31
        2.1   Equations and inequalities ......................................... 31
        2.2   Linear equations ................................................... 33
        2.3   Linear inequalities ................................................. 34
        2.4   Absolute-value equations and inequalities .......................... 37
        2.5   Functions ......................................................... 39
        2.6   Systems of linear equations ........................................ 42
        2.7   Systems of linear inequalities ...................................... 44
        2.8   Operations with exponents ......................................... 46
        2.9   Properties of logarithms ........................................... 47
        2.10  Solving problems involving exponential or logarithmic functions ..... 48
        2.11  Expanding polynomials ........................................... 48
        2.12  Factoring polynomials ............................................ 48
        2.13  Quadratic equations and quadratic expressions ................... 52
        2.14  Modeling nonlinear functions from real-world data ................ 53

2.15 Recursive patterns and relations ......................................... 54
2.16 Parametric functions.................................................. 56
2.17 Finding powers of a binomial (the binomial theorem)........................ 57
2.18 Composition of functions .............................................. 58
2.19 Inverses of functions.................................................. 58
2.20 Finding inverses of functions ........................................... 58
2.21 Operations with radicals............................................... 59
2.22 Piecewise functions .................................................. 60
2.23 Inverse variation..................................................... 61
2.24 Rational functions.................................................... 62
2.25 Rational expressions ................................................. 63
2.26 Polynomial functions.................................................. 66
2.27 The fundamental theorem of algebra .................................... 67
2.28 The factor theorem ................................................... 68
2.29 The rational root theorem.............................................. 68
2.30 The complex conjugate root theorem ................................... 69
2.31 Quadratic inequalities................................................. 69
2.32 Modeling functions ................................................... 70

**Chapter 3: Coordinate Geometry** ........................................... 73

3.1 Graphing linear equations ............................................. 73
3.2 Graphing quadratic equations ......................................... 75
3.3 Transformational geometry: translations, rotations, reflections, and scaling......... 80
3.4 Translations ........................................................ 80
3.5 Rotations .......................................................... 83
3.6 Reflections ......................................................... 84
3.7 Multiple transformations .............................................. 85
3.8 Symmetry.......................................................... 88
3.9 Polar coordinates .................................................... 91

**Chapter 4: Three-Dimensional Geometry** ................................... 93

4.1 Prisms............................................................. 93
4.2 Pyramids........................................................... 94
4.3 Curved figures ...................................................... 94
4.4 Similar solids and scale factors ........................................ 95
4.5 Three-dimensional coordinates ........................................ 96

**Chapter 5: Trigonometry** ................................................... 99

5.1 Trigonometric functions ............................................... 99
5.2 The Law of Sines .................................................... 109
5.3 The law of cosines ................................................... 110
5.4 Sum and difference formulas .......................................... 110
5.5 Double- and half-angle identities ....................................... 111
5.6 Proving trigonometric identities......................................... 112

**Chapter 6: Data Analysis, Statistics and Probability** ........................ 115

6.1 Factorials and their uses.............................................. 115
6.2 Measures of central tendency and dispersion in a dataset.................. 116
6.3 Displaying statistical data ............................................. 119

 6.4 Regression models. . . . . . . . . . . . . . . . . . . . . . . . . . . . . . . . . . . . . . . . . . . . . . . . 123
 6.5 Probability. . . . . . . . . . . . . . . . . . . . . . . . . . . . . . . . . . . . . . . . . . . . . . . . . . . . . . 127

## Section III: Practice Test 1 . . . . . . . . . . . . . . . . . . . . . . . . . . . . . . . . . . . . . . . . . . . . . . . . . 133

 Practice Test 1 . . . . . . . . . . . . . . . . . . . . . . . . . . . . . . . . . . . . . . . . . . . . . . . . . . . . . . . . 135

 Answer Key . . . . . . . . . . . . . . . . . . . . . . . . . . . . . . . . . . . . . . . . . . . . . . . . . . . . . . . . . 151

 Rationales. . . . . . . . . . . . . . . . . . . . . . . . . . . . . . . . . . . . . . . . . . . . . . . . . . . . . . . . . . . 153

## Section IV: Practice Test 2 . . . . . . . . . . . . . . . . . . . . . . . . . . . . . . . . . . . . . . . . . . . . . . . . 177

 Practice Test 2 . . . . . . . . . . . . . . . . . . . . . . . . . . . . . . . . . . . . . . . . . . . . . . . . . . . . . . . . 179

 Answer Key . . . . . . . . . . . . . . . . . . . . . . . . . . . . . . . . . . . . . . . . . . . . . . . . . . . . . . . . . 191

 Rationales . . . . . . . . . . . . . . . . . . . . . . . . . . . . . . . . . . . . . . . . . . . . . . . . . . . . . . . . . . 193

# SECTION I:
# Overview of the SAT Math Level 2 Test

# Overview

The SAT Math Level 2 exam tests your knowledge of about three years of college preparatory mathematics. This includes two years of algebra, one year of geometry. and some precalculus and trigonometry. The focus of the test is on problem solving, so you need to have your thinking hat on, and have an understanding of the concepts and their application to the problem. The Math Level 2 test covers more advanced skills in number and operations, algebra and functions, geometry and measurement, and data analysis, statistics and probability, than the Math Level 1 test.

The test consists of 50 multiple-choice questions; you have 60 minutes to complete the test.

In the 2015–2016 testing year, the SAT Math Level 2 Subject Area is offered six times a year. You cannot take the SAT and the SAT subject tests on the same date, so plan to take the Math Level 2 test in the next test date offering after you take the SAT. Some of the material for the SAT and the SAT Subject Area Math 2 section overlap, so, taking the Math Level 2 right after the SAT can be helpful.

## What Are The SAT Subject Area Tests?

The SAT Subject Area tests are designed to test your knowledge and problem-solving ability in greater depth in a specific subject. The College Board for SATs administers 20 subject area tests. The SAT Subject tests are additional admissions test that give you a chance to show your college admissions evaluators, that you are knowledgeable in a specific area. In particular, if you intend to major or minor in a subject area, then doing well on the subject area tests, showcases your credentials for the college.

For a list of all the subject areas go to www.sat.collegeboard.org.

## Choosing Between Math Level 1 and Math Level 2

Math Level 1 and 2 tests cover very similar topics, with one major difference. Math Level 2 includes test questions requiring precalculus and trigonometry understanding. If you did well, a B or better, in your math courses covering two years of algebra, one year of geometry, and a year of precalculus and/or trigonometry, then consider taking the Math Level 2 exam. The College Board's website, has more information on how to decide whether to take Math Level 1 or 2 at https://sat.collegeboard.org/practice/sat-subject-test-preparation/mathematics-level-2. Look at this for current updates.

Remember that the Math Level 2 test emphasizes more-advanced content. Talk to your peers or students that have recently taken the Math 1 or 2 tests to get an idea of the content or talk with your math teacher. **According to the College Board, you can choose to take either test on test day, regardless of what test you registered for.** Please confirm this from their website, as you get closer to your testing date.

## Areas Tested In SAT Math Level 2

The SAT Subject Area for Math Level 2 covers material taught in most high school math courses. This includes two years of Algebra, one year of Geometry, one year of pre-calculus and trigonometry. You may find that you don't know the answer to every single question, as you are not expected to have learned every topic on the test. But, as you prepare for the test, use the topic lists below to study, and ensure that you are somewhat knowledgeable about all the topics.

*Numbers and Operations*                                     *10%–14% (6 questions)*

Operations, ratio and proportion, complex numbers, counting, elementary number theory, matrices, sequences, series, vectors

*Algebra and Functions*                                      *48%–52% (25 questions)*

Expressions, equations, inequalities, representation and modeling, properties of functions (linear, polynomial, rational, exponential, logarithmic, trigonometric, inverse trigonometric, periodic, piecewise, recursive, parametric)

*Geometry and Measurement*                                   *28%–32% (13 questions)*

Coordinate: Lines, parabolas, circles, ellipses, hyperbolas, symmetry, transformations, polar coordinates                              10%–14%

Three-dimensional: Solids, surface area and volume (cylinders, cones, pyramids, spheres, prisms), coordinates in three dimensions        4%–6%

Trigonometry: Right triangles, identities, radian measure, law of cosines, law of sines, equations, double angle formulas              12%–16%

*Data Analysis, Statistics and Probability*                  *8%–12% (6 questions)*

Mean, median, mode, range, interquartile range, standard deviation, graphs and plots, least squares regression (linear, quadratic, exponential), probability

## Why Take The SAT Math Level 2 Test?

The SAT Subject test is an opportunity to show the colleges you are applying to that you know more about and better understand a specific subject area. It is not required that you take the SAT Subject Area tests for admission to college, however, if you have taken the three years of high school math in algebra, geometry, precalculus and trigonometry, are comfortable with the subject matter, and did reasonably well in the coursework, then, it might be an easy test to take to add to your achievements.

While most colleges do not REQUIRE the Subject Area tests, several of them CONSIDER it for admissions and placement. Take a look at the test requirements for the colleges you are interested in applying to, and then decide on which SAT Subject tests will enhance your application. Another reason to consider the Subject Area test is that sometimes, you might be able to waive an introductory math course, depending on your scores on the Math test. In other cases, the SAT Subject Area test is considered as a replacement for an ACT test. So, as you get ready in your junior year to take the SAT, look at the admissions requirements for the colleges you might be interested in applying to, and see if there is a benefit to taking the Math Level 2 test to highlight your knowledge in math.

## How Is The SAT Math Level 2 Test Scored?

The total score you can get for Math Level 2 is 800. The score range is from 200 to 800, and increases in 20-point intervals. **According to the College Board, you do not have to get every question correct to receive the highest score for the test.** You will receive 1 point for each question you answer correctly. It's important to understand the ramifications of guessing on this test. If you answer a question incorrectly, then points are subtracted, based on the following scale:
- ¼ point subtracted for each 5-choice question
- ⅓ point subtracted for each 4-choice question
- ½ point subtracted for each 3-choice question
- 0 points subtracted for questions you don't answer

The raw scores are then equated to the 200–800 point scale. This ensures that different tests and scores of other students do not affect your score. It also means that you score is not graded on a curve (where the highest scorer gets 800, and the remaining get scores relative to that).

Three types of scores are reported for the test you took. (1) Your score on the 200–800 point range. (2) The average score based on the most recent tests in that subject area. (3) The percentile score—for example, for your score of 700, if your percentile is 85, then it means that you did better than 85% of the students taking that test.

In the year 2014, below you see the percentiles and corresponding scores for the Math Level 2 test.[1] So, 51 percent of the students scored below 710 and 49 percent of the students scored above 710. As you see, the mean score for the Math Level 2 is very high—over 700.
- 800   81 percentile
- 780   74 percentile
- 710   51 percentile
- 630   24 percentile

Cancelling Scores: If you decide after taking the test that you want to cancel your score, you can do so immediately on test day. However, the SAT College Board warns that it will cancel all your scores for the day.

## Should You Guess Answers on the Test?

Now that you know that your score is penalized for incorrect guessing, you should evaluate on a question-by-question basis whether to guess the answer. If you have some

---

[1] Percentile Scores for SAT Subject Area tests administered in 2014. https://secure-media.collegeboard.org/digitalServices/pdf/sat/sat-percentile-ranks-subject-tests-2014.pdf

understanding of the question being asked and the subject matter and can discard one or two of the multiple-choice responses, then take an EDUCATED guess. It's probably not worth losing points by WILD GUESSING. The best option is to mark that question on the test sheet and go back to it once you've finished the test.

## How Is The SAT Math Level 2 Exam Administered?

The SAT subject area tests are administered by the College Board, the same organization that administers the SAT exams. In the 2015–2016 year, the Math Level 2 test is given six times: October, November, December, January, February, and June. So if you are a junior, you should plan on taking the SAT first and then taking the Math Level 2 test at the next date window. So for example, if you take the SAT in October, you can then take the Math Level 2 test in November. As mentioned earlier, there is some similarity in the math subject matter between the SAT and Math Level 2. So if you can, take the Math Level 2 test while the SAT material is still fresh.

You will find registration information online at www.satcollegeboard.org. From the website you can register online or by mail. In both cases, you will need to upload a digital photo of yourself as identification. You can find more about the photo requirements at https://sat.collegeboard.org/register/photo-requirements. The photo will become part of your exam admission ticket. You get four registration score reports for the fee (so think about the four colleges you want to send your scores to), additional reports will need to be paid for separately.

## Accommodations for Students with Disabilities

If you have a documented disability, you may be eligible for special accommodations for the test. The SAT College Board's website describes the requirements to obtain the approvals for Services for Students with Disabilities (SSD). Look up the current requirements and steps to register at: https://sat.collegeboard.org/register/for-students-with-disabilities

In order to obtain the SSD approvals, you will need to file a request with the SAT College Board, well before you intend to take the test. The approvals take about seven weeks to process, and documentation of the student's disability and need for specific accommodations is required. The deadlines for SSD approvals are much earlier than the SAT Subject Area deadline. If you intend to take the SAT (and the SAT Subject Area) in the fall of your junior year, then the College Board recommends applying for your accommodation approvals in the spring of your sophomore year. Keep in mind, that if you were approved for special accommodations by the College Board for the SAT, then you might not have to reapply for the Math Level 2 test. But, it is always helpful, to confirm this with the College Board.

Once you are approved for special accommodation for test-taking, the accommodation will be noted on your admission ticket. If your accommodation request is not approved, then you must take the test as a standard test-taker.

## Getting Ready for Test Day

Prepare your test day material, the day before. The tests usually start in the morning, so don't wait until the last minute to find your gear. Here's what you will need at a minimum:

1) A printed out copy of your Admission Ticket
2) Photo identification
3) No. 2 pencils and an eraser
4) An approved calculator
5) A watch, so you know how much time is remaining

*Calculator:* Take a calculator that you are comfortable using, with you on test day. Before you pick up a calculator for your test, think about the problem solution first. You might find that you only need the calculator for the last step or two. You will need a calculator for about 55%–65% of the questions for the Math Level 2 test. Remember, when doing a calculation, DO NOT ROUND intermediate calculations; carry the results forward, until the final answer.

You can take either a graphing calculator or a scientific calculator with you. The College Board's website lists the acceptable calculators, and recommends a graphing calculator over a scientific calculator.

You will not be allowed to use the calculator function on devices such as; a laptop, tablet, cell phone or smart phone. Calculator models that access the Internet, have wireless or Bluetooth capabilities will not be allowed. Models that have a typewriter-like keypad, pen or stylus input are also not allowed. Check to make sure you have new batteries in your calculator, or know that it will work for the duration of the test. You will not be able to borrow another test-taker's calculator during the test, in case your calculator fails. Check the latest list of acceptable calculators on the College Board's website, before your test day.

You cannot bring the following into the testing room:

1) Any form of cellphone, tablet, or computer device
2) iPods or music devices
3) Cameras or recording devices

*Geometric Figures:* When you are solving a math problem with a figure, assume that the figure is drawn to scale, unless it specifically states that it is not to scale. So, what does this mean? This means that the relative positions of points and angles are in the order shown. And line segments that extend through lines, and appear to lie on the same line, are assumed to be on the line.

*Reference Information Provided on the Test:* The formulas below are provided on the Math Level 2 test. So, you don't have to memorize them.

- Volume of a right circular cone with radius $r$ and height $h$: $V = \frac{1}{3}\pi r^2 h$
- Volume of a sphere with radius $r$: $V = \frac{4}{3}\pi r^3$
- Volume of a pyramid with base area $B$ and height $h$: $V = \frac{1}{3}Bh$
- Surface area of a sphere with radius $r$: $S = 4\pi r^2$

## Testing Tips

1. **Get smart, play dumb.** Sometimes a question is just a question. No one is out to trick you, so don't assume that the test writer is looking for something other than what was asked. Stick to the question as written and don't overanalyze.

2. **Do a double-take.** Read test questions and answer choices at least twice because it's easy to miss something, to transpose a word or some letters. If you have no idea what the correct answer is, skip it and come back later if there's time.

3. **Turn it on its ear.** The syntax of a question can often provide a clue, so make things interesting and turn the question into a statement to see if it changes the meaning or relates better (or worse) to the answer choices.

4. **Get out your magnifying glass.** Look for hidden clues in the questions, because it's difficult to write a multiple-choice question without giving away part of the answer in the options presented. In most questions you can readily eliminate one or two potential answers, increasing your chances of answering correctly to 50/50, which will help out if you've skipped a question and gone back to it (see tip #2). So, read the question carefully.

5. **Call it intuition.** Often your first instinct is correct. If you've been studying the content you've likely absorbed something and have subconsciously retained the knowledge. On questions you're not sure about trust your instincts, because a first impression is usually correct.

6. **Graffiti.** Sometimes it's a good idea to mark your answers directly on the test booklet and go back to fill in the optical scan sheet later. You don't get extra points for perfectly blackened ovals. If you choose to manage your test this way, be sure not to mismark your answers when you transcribe to the scan sheet.

7. **Become a clock-watcher.** You have a set amount of time to answer the questions. Don't get bogged down laboring over a question you're not sure about when there are ten others you could answer more readily. If you choose to follow the advice of tip #6, be sure you leave time near the end to go back and fill in the scan sheet.

# SECTION II:
# Content Review for SAT Math Level 2

# Chapter 1: Number and Operations

## 1.1 Order of operations

When simplifying algebraic expressions we use the following order:
1. Perform operations within a parenthesis.
2. Evaluate exponents.
3. Multiply and divide from left to right.
4. Add and subtract from left to right.

**Example:**

$$3 + 2(4 + 3)^2 - 10 \div 5 = 3 + 2(7)^2 - 10 \div 5 \quad \text{Perform operations in parentheses.}$$
$$= 3 + 2(49) - 10 \div 5 \quad \text{Evaluate exponents.}$$
$$= 3 + 98 - 2 \quad \text{Multiply and divide from left to right.}$$
$$= 99 \quad \text{Add and subtract from left to right.}$$

## 1.2 Properties of operations

**Properties of operations** are rules that apply for addition, subtraction, multiplication, or division of real numbers.

### Commutative Property

You can change the order of the terms or factors as follows.

For addition: $a + b = b + a$
For multiplication: $ab = ba$

This rule does not apply for division and subtraction.

**Examples:** $17 + 23 = 23 + 17 = 40$
$8 \times 19 = 19 \times 8 = 152$

### Associative Property

You can regroup the terms as you like.

For addition: $a + (b + c) = (a + b) + c$
For multiplication: $a(bc) = (ab)c$

This rule also does not apply for division and subtraction.

**Example:** $(-2 + 7) + 5 = -2 + (7 + 5)$
$5 + 5 = -2 + 12 = 10$

**Example:** $(3 \times -7) \times 5 = 3 \times (-7 \times 5)$
$-21 \times 5 = 3 \times -35 = -105$

## Identity Properties

Adding 0 to a number results in that number, with no change (additive identity of 0); multiplying a number by 1 results in that number, with no change (multiplicative identity of 1).

For addition: $a + 0 = a$ (additive identity of 0)
For multiplication: $a \times 1 = a$ (multiplicative identity of 1)

**Example:** $17 + 0 = 17$

**Example:** $-34 \times 1 = -34$

## Inverse Properties

The additive inverse of a number $a$ is the number that when added to $a$ results in zero. The multiplicative inverse (or reciprocal) of a number a is the number that when multiplied by $a$ results in 1.

For addition: $a + (-a) = 0$

**Example:** $25 + -25 = 0$

For multiplication: $a \times \frac{1}{a} = 1$ (multiplicative identity of 1)

**Example:** $5 \times \frac{1}{5} = 1$

The additive inverse of $a$ is $(-a)$. The reciprocal or multiplicative inverse of $a$ is $\frac{1}{a}$.

The sum of any number and its additive inverse is 0. The product of any number and its reciprocal is 1.

## Distributive Property of Multiplication over Addition and Subtraction

This property allows us to operate on terms within parentheses without first performing operations within the parentheses. This is especially helpful when terms within the parentheses cannot be combined.

$a(b + c) = ab + ac$

**Example:** $6 \times (-4 + 9) = (6 \times -4) + (6 \times 9)$
$6 \times 5 = -24 + 54 = 30$

To multiply a sum by a number, multiply each addend by the number, then add the products.

**Summary of the properties of operations**

| Property | of Addition | of Multiplication |
|---|---|---|
| Commutative | $a + b = b + a$ | $ab = ba$ |
| Associative | $a + (b + c) = (a + b) + c$ | $a(bc) = (ab)c$ |
| Identity | $a + 0 = a$ | $a \times 1 = a$ |
| Inverse | $a + (-a) = 0$ | $a \times \dfrac{1}{a} = 1,\ a \neq 0$ |
| Distributive property of multiplication over addition and subtraction | $a(b + c) = ab + ac$ | $a(b - c) = ab - ac$ |

## 1.3 Ratios

A **ratio** is a comparison of two numbers for the purpose of relating relative magnitudes. For instance, if a class had 11 boys and 14 girls, the ratio of boys to girls could be written one of three ways:

$$11 : 14 \text{ or } 11 \text{ to } 14 \text{ or } \frac{11}{14}.$$

The ratio of girls to boys is:

$$14 : 11 \text{ or } 14 \text{ to } 11 \text{ or } \frac{14}{11}.$$

Ratios, like fractions, can be reduced if their terms have common factors. A ratio of 12 cats to 18 dogs, for example, reduces to $2 : 3$, 2 to 3 or $\frac{2}{3}$.

## 1.4 Proportions

A **proportion** is an equation in which a fraction is set equal to another fraction. To solve the proportion, multiply each numerator times the other fraction's denominator. Set these two products equal to each other and solve the resulting equation. This is called cross-multiplying the proportion.

**Example:** Find $x$ given the proportion $\dfrac{4}{15} = \dfrac{x}{60}$.

To solve for $x$, cross-multiply.

$$(4)(60) = (15)(x)$$
$$240 = 15x$$
$$16 = x$$

**Example:** Find $x$ given the proportion $\dfrac{x+3}{3x+4} = \dfrac{2}{5}$.

To solve for $x$, cross-multiply.

$$5(x+3) = 2(3x+4)$$
$$5x + 15 = 6x + 8$$
$$7 = x$$

The mathematics of solving for variables in proportions is not difficult. The key to solving a problem involving proportions is constructing the proportion correctly. As noted above, this requires carefully reading the problem, followed by careful identification of the related values and construction of the appropriate ratios.

Proportions can be used to solve word problems whenever relationships are compared. Some situations include scale drawings and maps, similar polygons, speed, time and distance, cost, and comparison shopping.

**Example:** Which is the better buy, 6 items for $1.29 or 8 items for $1.69?

Find the unit price.

$$\dfrac{6}{1.29} = \dfrac{1}{x} \qquad \dfrac{8}{1.69} = \dfrac{1}{x}$$
$$6x = 1.29 \qquad 8x = 1.69$$
$$x = 0.215 \qquad x = 0.21125$$

Thus, 8 items for $1.69 is the better buy.

**Example:** A car travels 125 miles in 2.5 hours. How far will it go in 6 hours?

Write a proportion comparing distance and time.

$$\dfrac{miles}{hours} : \quad \dfrac{125}{2.5} = \dfrac{x}{6}$$
$$2.5x = 750$$
$$x = 300$$

The car will travel 300 miles in 6 hours.

**Example:** The scale on a map is $\dfrac{3}{4}$ inch = 6 miles. What is the actual distance between two cities if they are $1\dfrac{1}{2}$ inches apart on the map?

Write a proportion comparing the scale to the actual distance.

$$\overset{\text{scale}}{\dfrac{\frac{3}{4}}{1\frac{1}{2}}} = \overset{\text{actual}}{\dfrac{6}{x}}$$

$$\frac{3}{4}x = 1\frac{1}{2} \times 6$$

$$\frac{3}{4}x = 9$$

$$x = 12$$

Thus, the actual distance between the cities is 12 miles.

## 1.5 Natural (counting) numbers

The set of **natural numbers**, $\mathbb{N}$, includes 1, 2, 3, 4, … (By some definitions, $\mathbb{N}$ includes zero.) The natural numbers are sometimes called the counting numbers (especially if the definition of $\mathbb{N}$ excludes zero).

The set $\mathbb{N}$ obeys the properties of associativity, commutativity, distributivity and identity for multiplication and addition (assuming, for the case of addition, that zero is included in some sense in the natural numbers). The set of natural numbers does not contain additive or multiplicative inverses, however, as there are no noninteger fractions or negative numbers.

Natural numbers can be either even or odd. **Even** numbers are evenly divisible by two; **odd** numbers are not evenly divisible by two (alternatively, they leave a remainder of one when divided by two).

## 1.6 Properties of odd and even numbers

### Addition

| Rule | Example |
|---|---|
| Odd + Odd = Even | 5 + 7 = 12 |
| Odd + Even = Odd | 9 + 22 = 31 |
| Even + Odd = Odd | 16 + 13 = 29 |
| Even + Even = Even | 28 + 4 = 32 |

### Multiplication

| Rule | Example |
|---|---|
| Odd × Odd = Odd | 5 × 7 = 35 |
| Odd × Even = Even | 9 × 22 = 198 |
| Even × Odd = Even | 16 × 13 = 208 |
| Even × Even = Even | 28 × 4 = 112 |

## 1.7 Composite and prime numbers

Any natural number $n$ that is divisible by at least one number that is not equal to 1 or $n$ is called a **composite number**. A **prime number** is a natural number $n$ that is divisible by two numbers only: 1 and $n$. The number 1 does not qualify as a prime number because it is divisible by only one number, itself, whereas a prime is divisible by exactly two numbers.

**Example:** 91 is divisible by 7 and 13 in addition to being divisible by 91 and 1, so 91 is a composite number.

93 is divisible only by 93 and 1, so 93 is a prime number.

## 1.8 Divisibility tests for natural numbers

In many cases, it is possible to say whether a natural number is divisible by a certain factor without actually doing the division, by applying a divisibility test. There is such a test for every number from 1 to 12 except 7.

1. Every natural number is divisible by 1.
2. A number is divisible by 2 if that number is an even number (i.e., the last digit is 0, 2, 4, 6 or 8).

    Consider a number defined by the digits $a$, $b$, $c$ and $d$ (for instance, 1,234). Rewrite the number as follows. The value of the number is
    $$1000a + 100b + 10c + d$$
    The first three terms are divisible by 2. Thus, the number is only divisible by 2 if $d$ is divisible by two.

    For example, the last digit of 1,354 is 4, so it is divisible by 2. On the other hand, the last digit of 240,685 is 5, so it is not divisible by 2.

3. A number is divisible by 3 if the sum of its digits is evenly divisible by 3. Consider a number defined by the digits $a$, $b$, $c$ and $d$. The number can be written as
    $$1000a + 100b + 10c + d$$
    The number can also be rewritten as
    $$(999 + 1)a + (99 + 1)b + (9 + 1)c + d$$
    $$999a + 99b + 9c + (a + b + c + d)$$
    Note that the first three terms in the above expression are all divisible by 3. Thus, the number is evenly divisible by 3 only if $a + b + c + d$ is divisible by 3. The same logic applies regardless of the size of the number. This proves the rules for divisibility by 3.

    The sum of the digits of 964 is $9 + 6 + 4 = 19$. Since 19 is not divisible by 3, neither is 964. The sum of the digits of 86,514 is $8 + 6 + 5 + 1 + 4 = 24$. Since 24 is divisible by 3, 86,514 is also divisible by 3.

4. A number is divisible by 4 if the last two digits of the number are evenly divisible by 4.

    Let a number $abcd$ be defined by the digits $a$, $b$, $c$ and $d$.
    $$1000a + 100b + 10c + d$$
    The number can also be written as $100(10a + b) + 10c + d$

Since 100 is divisible by 4, $100(10a + b)$ is also divisible by 4. Thus, *abcd* is divisible by 4 only if the two-digit number *cd* is divisible by 4.

The number 113,336 ends with the number 36 for the last two digits. Since 36 is divisible by 4, 113,336 is also divisible by 4. The number 135,626 ends with the number 26 for the last two digits. Since 26 is not evenly divisible by 4, 135,626 is also not divisible by 4.

5. A number is divisible by 5 if the number ends in either a 5 or a 0. Once again, imagine a number with the digits *abcd*. The number equals

    $$1000a + 100b + 10c + d$$

    The first three terms are evenly divisible by 5, but the last term is only evenly divisible by 5 if *d* is divisible by 5, that is, if the number ends in a 0 or a 5. For instance, 225 ends with a 5, so it is divisible by 5. The number 470 is also divisible by 5 because its last digit is a 0. The number 2,358 is not divisible by 5 because its last digit is an 8.

6. A number is divisible by 6 if the number is divisible by both 2 and 3. Thus, any even number that is divisible by 3 is also divisible by 6. For instance, 4,950 is an even number and its digits add up to $4 + 9 + 5 + 0 = 18$. Since it is even and the sum of its digits is divisible by 3, the number 4,950 is divisible by 3 and by 6 as well. On the other hand, 326 is an even number, but its digits add up to 11, which is not divisible by 3. Therefore, 326 is not divisible by 3 or by 6.

7. There is no easy test of divisibility by 7.

8. A number is divisible by 8 if the number in its last three digits is evenly divisible by 8. The logic for the proof of this case follows that of numbers divisible by 2 and 4. Taking a number with digits *abcd*, we can write its value as

    $$1000a + 100b + 10c + d$$

    The first term is divisible by 8, so the number as a whole is divisible by 8 only if the last three terms compose a number divisible by 8.

    **Example:** The number 113,336 ends with the 3-digit number 336. Since 336 is divisible by 8, then 113,336 is also divisible by 8. The number 465,628 ends with the 3-digit number 628. Since 628 is not evenly divisible by 8, then 465,628 is also not divisible by 8.

9. A number is divisible by 9 if the sum of its digits is evenly divisible by 9. The logic for the proof of this case follows that for the case of numbers that are divisible by 3 and 6. The sum of the digits of 874, for example, is $8 + 7 + 4 = 19$. Since 19 is not divisible by 9, neither is 874. The sum of the digits of 116,514 is $1 + 1 + 6 + 5 + 1 + 4 = 18$. Since 18 is divisible by 9, 116,514 is also divisible by 9.

10. A number is divisible by 10 if the last digit is zero.

11. A number is divisible by 11 if the sum of every other digit differs from the sum of the intervening digits by 0 or a multiple of 11.

**Example:** The sum of every other digit of 132,847 is 1 + 2 + 4 = 7. The sum of the intervening digits is 3 + 8 + 7 = 18. The difference of 18 and 7 is 11, so 132,847 is divisible by 11.

The sum of every other digit of 61,955 is 6 + 9 + 5 = 20. The sum of the intervening digits is 1 + 5 = 6. The difference of 20 and 6 is 14, so 61,955 is not divisible by 11.

12. A number is divisible by 12 if it is divisible by both 4 and 3: the last two digits are divisible by 4 and the sum of the digits is divisible by 3.

**Example:** The last two digits of 864 are 64, a number divisible by 4. The sum of the digits is 8 + 6 + 4 = 18, which is divisible by 3. Therefore, 864 is divisible by 12.

The strategies used in these divisibility tests can be extended to other higher factors. A number is divisible by 15, for instance, if it is divisible by 3 and 5, and it is divisible by 44 if it is divisible by 4 and 11.

## 1.9 Real numbers

The following chart shows the relationships among the subsets of the real numbers.

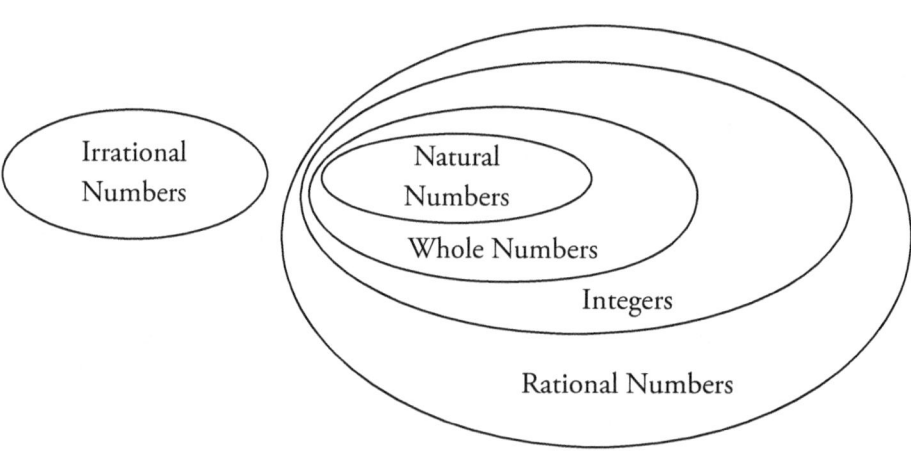

**Real numbers** are denoted by $\mathbb{R}$ and are numbers that can be shown by an infinite decimal representation such as 3.286275347 . . . . Real numbers include **rational numbers**, numbers that can be represented as the quotient of two integers, such as 242 and $-\frac{23}{129}$, and **irrational numbers**, such as $\sqrt{2}$ and $\pi$, which cannot. All real numbers can be represented as points along an infinite number line.

Real numbers are to be distinguished from imaginary numbers.

Real numbers are classified as follows:

| | |
|---|---|
| **Natural Numbers Denoted by** $\mathbb{N}$ | The counting numbers. 1, 2, 3, . . . |
| **Whole Numbers** | The counting numbers along with zero. 0, 1, 2, 3, . . . |
| **Integers Denoted by** $\mathbb{Z}$ | The counting numbers, their negatives, and zero. . . . , −2, −1, 0, 1, 2, . . . |
| **Rationals, Denoted by** $\mathbb{Q}$ | All of the fractions that can be formed using whole numbers. Zero cannot be the denominator. In decimal form, these numbers will be either terminating or repeating decimals. Simplify square roots to determine if the number can be written as a fraction. |
| **Irrationals** | Real numbers that cannot be written as a fraction. The decimal forms of these numbers neither terminate nor repeat. Examples include $\pi$, $e$, and $\sqrt{2}$ |

## 1.10 Matrices

A **matrix** is an ordered set of numbers written in rectangular form. An example matrix is shown below.

$$\begin{bmatrix} 0 & 3 & 1 \\ 4 & 2 & 3 \\ 1 & 0 & 2 \end{bmatrix}$$

Since this matrix has 3 rows and 3 columns, it is called a $3 \times 3$ matrix. If we named this matrix $A$, the element in the second row of the third column would be denoted as $A_{2,3}$. In general, a matrix with $r$ rows and $c$ columns is an $r \times c$ matrix.

Matrix addition and subtraction obey the rules of commutativity, associativity, identity, and additive inverse.

$$A + B = B + A$$

$$A + (B + C) = (A + B) + C$$

$$A + 0 = A$$

$$A + (-A) = 0$$

### Addition and subtraction of matrices

Matrices can be added or subtracted only if their dimensions are the same. To add or subtract compatible matrices, simply add or subtract the corresponding elements.

Example: $\begin{bmatrix} 2 & 4 \\ 0 & 5 \end{bmatrix} + \begin{bmatrix} 1 & 0 \\ 7 & 8 \end{bmatrix} = \begin{bmatrix} 2+1 & 4+0 \\ 0+7 & 5+8 \end{bmatrix} = \begin{bmatrix} 3 & 4 \\ 7 & 13 \end{bmatrix}$

## Multiplication of matrices

Matrices can be multiplied by a scalar, that is, a single number. The product of a matrix and a scalar is found by multiplying each element of the matrix by the scalar.

**Example:** $7\begin{bmatrix} 4 & 1 \\ 8 & 0 \end{bmatrix} = \begin{bmatrix} 28 & 7 \\ 56 & 0 \end{bmatrix}$

Two matrices can be multiplied by each other if the second matrix has as many rows as the first matrix has columns. To put it another way, an $a \times b$ matrix can be multiplied by a $b \times c$ matrix, where $b$ is the same number in both cases. The result is an $a \times c$ matrix.

To perform the multiplication, each row of the first matrix is multiplied by each column of the second matrix as follows: The first item in the row is multiplied by the first item in the column; the second item in the row is multiplied by the second item in the column, and so on, and the products are added. Multiplying row $m$ in the first matrix by column $n$ in the second matrix produces item $x_{m,n}$ in the product matrix.

**Example:** matrix $A = \begin{bmatrix} 1 & 2 & 6 \\ 4 & 5 & 0 \end{bmatrix}$ matrix $B = \begin{bmatrix} 7 & 1 \\ 8 & 0 \\ 3 & 3 \end{bmatrix}$. Find matrix $C = AB$.

$A$ is a $2 \times 3$ matrix, while $B$ is a $3 \times 2$ matrix. $C$ will be a $2 \times 2$ matrix.

Multiply row 1 of matrix $A$ by column 1 of matrix $B$ to find $C_{1,1}$:

$$C_{1,1} = 1 \times 7 + 2 \times 8 + 6 \times 3 = 7 + 16 + 18 = 41$$

Multiply row 1 of matrix $A$ by column 2 of matrix $B$ to find $C_{1,2}$:

$$C_{1,2} = 1 \times 1 + 2 \times 0 + 6 \times 3 = 1 + 0 + 18 = 19$$

Similarly, multiply row 2 of matrix $A$ by column 1 of matrix $B$ to find $C_{2,1}$ and multiply row 2 of matrix $A$ by column 2 of matrix $B$ to find $C_{2,2}$, completing the matrix:

$$C = \begin{bmatrix} 41 & 19 \\ 68 & 4 \end{bmatrix}$$

## Determinants of matrices

Associated with every square matrix is a number called its **determinant**. The determinant of a matrix is typically denoted using straight brackets; thus, the determinant of matrix $A$ is $|A|$. Use the following formula to calculate the determinant of a $2 \times 2$ matrix:

$$\begin{vmatrix} a & b \\ c & d \end{vmatrix} = ad - bc$$

Example: find the determinant of

$$\begin{bmatrix} 4 & -8 \\ 7 & 3 \end{bmatrix}$$

Use the formula for calculating the determinant.

$$\begin{vmatrix} 4 & -8 \\ 7 & 3 \end{vmatrix} = (4 \times 3) - (-8 \times 7) = 12 - (-56) = 68$$

## 1.11 Vectors

A **vector** is defined by having both magnitude and direction. Examples would be a force of 10 pounds pushing downward or a train traveling 80 miles an hour due east. Vectors are typically represented by an arrow; the angle of the arrow shows its direction, and the length of the arrow shows its magnitude.

Example: A wind is blowing from the east at 3 mph. Represent it by a vector.

If the wind is from the east, it is blowing westward. That can be represented by an arrow pointing left. The speed of 3 mph can be represented by an arrow 3 units long.

Vectors can also be represented by a pair of coordinates. The magnitude of such a vector is the length of the line drawn from the origin to the point indicated by the coordinates. It can be found using the Pythagorean Theorem.

Example: Find the magnitude $r$ of the vector $(3,-4)$.

$$r = \sqrt{3^2 + (-4)^2} = \sqrt{9+16} = 5$$

The vector has a magnitude of 5.

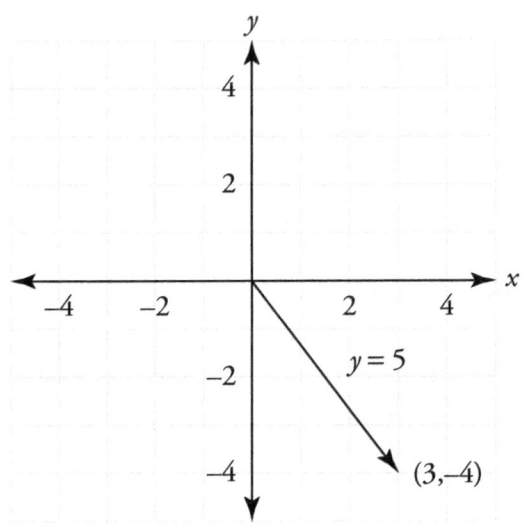

**Addition of vectors**

Vectors indicated by coordinates can be added by simply adding each of the coordinates for the two vectors.

**Example:** $(-5, 12) + (7, -2) = (2, 10)$.

Vectors can be added graphically by placing the tail of the second arrow at the head of the first and drawing an arrow from the tail of the first arrow to the head of the second:

**Example:** A canoe paddles northward at 2 mph in a current running eastward at 2 mph. What is the total speed and direction of the canoe?

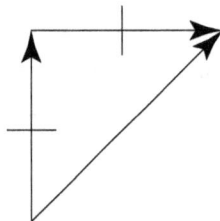

From the diagram we can see that the magnitude, in this case the total speed, is $\sqrt{2^2 + 2^2} = \sqrt{8} = 2\sqrt{2} \approx 2.83$ mph. The diagram also shows that the direction is northeast.

**Subtraction of vectors**

Vectors indicated by coordinates can be subtracted simply by subtracting the two sets of coordinates.

**Example:** $(4, 8) - (5, -12) = (-1, 20)$

Vectors can also be subtracted graphically. The second vector, the vector to be subtracted, is placed with its tail at the head of the first vector, the vector being subtracted from. Then a vector of the same length as the second vector, but in the opposite direction, is drawn from the head of the first vector. Finally, a vector is drawn from the tail of the first vector to the head of the reversed vector.

**Example:** Subtract vector **b** from vector **a**.

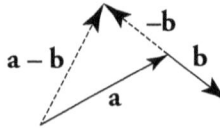

## 1.12  Sequences and series

Sequences and series can take on a vast range of different forms and patterns. Sequences and series are essentially two different representations of a set of numbers: a **sequence** is the set of numbers, and a **series** is the sum of the terms of the sequence. That is, a sequence such as

$a_1, a_2, a_3, \ldots$

has a corresponding series S such that
$$S = a_1 + a_2 + a_3 + \ldots$$
Two of the most common forms of series are the arithmetic and geometric series, both of which are discussed below.

## Arithmetic series

A finite series of numbers for which the difference between successive terms is constant is called an **arithmetic series**. An arithmetic series with $n$ terms can be expressed as follows, where $a$ and $d$ are constants. (The constant $a$ is the first term, and $d$ is the difference between successive terms.)

$$a + (a+d) + (a+2d) + (a+3d) + \ldots (a+[n-1]d)$$

To derive the general formula, examine the series sum for several small values of $n$.

| n | sum |
|---|---|
| 1 | $a$ |
| 2 | $2a + d$ |
| 3 | $3a + 3d$ |
| 4 | $4a + 6d$ |
| 5 | $5a + 10d$ |
| 6 | $6a + 15d$ |
| $\vdots$ | $\vdots$ |
| $n$ | $na + d\sum_{i=1}^{n-1} i$ |

The result in the table for $n$ terms is found by examining the pattern of the previous series. All that is necessary, then, is to determine a closed expression for the summation.

By inspection, it can be seen that the product of $n$ and $(n + 1)$, divided by 2, is the expression for the sum of $1 + 2 + 3 + 4 + 5 + \ldots + n$. Then:

$$\sum_{i=1}^{n} i = \frac{1}{2} n(n+1)$$

A simple derivation of this relationship may be made as follows:
$$S_n = 1 + 2 + 3 + \ldots + n$$
Writing the terms in reverse order:
$$S_n = n + (n-1) + (n-2) + \ldots + 1$$
Adding the two expressions for $S_n$ term by term, we get
$$2S_n = (1+n) + (2+n-1) + (3+n-2) + \ldots + (n+1)$$
$$= (1+n) + (1+n) + (1+n) + \ldots + (n+1)$$
$$= n(n+1)$$

Therefore, $S_n = \dfrac{n(n+1)}{2}$

For the general case (with first term $a$ and common difference $d$), therefore, the sum for a series with $n$ terms is given by

$$na + d\sum_{i=1}^{n-1} i = na + d\frac{(n-1)(n)}{2} = \frac{1}{2}n(2a + d(n-1))$$

Often, closed formulas for series such as the arithmetic series must be found by inspection, as a more rigorous derivation is difficult. The result can be proven using mathematical induction, however.

**Example:** Calculate the sum of the series $1 + 5 + 9 + \ldots + 57$.

This is an arithmetic series, as the difference between successive terms, $d$, is constant ($d = 4$). Determine the total number of terms by subtracting the first term from the last term, dividing by $d$, and adding 1.

$$n = \frac{57-1}{4} + 1 = \frac{56}{4} + 1 = 14 + 1 = 15$$

That this approach works can be seen by testing simple examples. For instance, if the series is $1 + 5 + 9$, then

$$n = \frac{9-1}{4} + 1 = 3$$

There are indeed three terms in this simple series. Next, apply the formula, noting that $a = 1$.

$$\frac{1}{2}n[2a + d(n1)] = \frac{1}{2}(15)[2(1) + 4(15-1)]$$
$$= \frac{15}{2}[2 + 4(14)] = \frac{15}{2}(58) = 435$$

Thus, the answer is 435.

### Geometric series

A **geometric series** is a series whose successive terms are related by a common factor (rather than the common difference of the arithmetic series). Assuming that $a$ is the first term of the series and $r$ is the common factor, the general $n$-term geometric series can be written as follows.

$$a + ar + ar^2 + ar^3 + \ldots + ar^{n-1}$$

The geometric series can also be written using sum notation.

$$a + ar + ar^2 + ar^3 + \ldots + ar^{n-1} = \sum_{i=0}^{n-1} ar^i$$

To derive the closed-form expression for this finite series, let the sum for $n$ terms be defined as $S_n$. Multiply $S_n$ by $r$.

$$S_n = a + ar + ar^2 + \ldots + ar^{n-1}$$
$$rS_n = ar + ar^2 + ar^3 + \ldots + ar^n$$

Note that if $a$ is added to this new series, the result is the sum $S_{n+1}$, which has $n + 1$ terms.

$$a + rS_n = a + ar + ar^2 + ar^3 + \ldots + ar^n = S_{n+1}$$

But $S_{n+1}$ is simply $S_n + ar^n$, so the above expression can be written solely in terms of $S_n$.

$$a + rS_n = S_{n+1} = S_n + ar^n$$

Rearrange the result to obtain a simple formula for the geometric series.

$$a + rS_n = S_n + ar^n$$
$$a - ar^n = S_n - rS_n$$
$$a(1 - r^n) = S_n(1 - r)$$
$$S_n = a\left(\frac{1 - r^n}{1 - r}\right)$$

Example: find the sum of the first 5 terms of a geometric sequence whose first term is 4 and whose common factor is 3.

$$a = 4,\ r = 3,\ n = 5$$

$$S_n = a\left(\frac{1 - r^n}{1 - r}\right) = 4\left(\frac{1 - 3^5}{1 - 3}\right) = 4\frac{-242}{-2} = 4(121) = 484$$

Check: $4 + 12 + 36 + 108 + 324 = 484$ ✓

## 1.13 Imaginary numbers

Sometimes, when solving an equation, you need to take the square root of a negative number. If so, the equation has no real-number solutions: since the square of either a positive or negative number is positive, no real number could be the square root of a negative number. To deal with the square roots of negative numbers, mathematicians have created the concept of $i$, the square root of $-1$. The square roots of other negative numbers can be given in terms of $i$, using standard operations with radicals: $\sqrt{-25} = \sqrt{(25)(-1)} = \sqrt{25}\left(\sqrt{-1}\right) = 5i$. Such numbers are called **imaginary numbers**.

Example: solve for $x$:

$$x^2 + 36 = 0$$
$$x^2 = -36$$
$$x = \pm 6i$$
$$x_1 = 6i,\ x_2 = -6i$$

## 1.14 Complex numbers

The sum of a real number and an imaginary number is considered a single quantity; numbers consisting of a real number plus or minus an imaginary number are called **complex numbers**.

The set of complex numbers is denoted by $\mathbb{C}$. The set $\mathbb{C}$ is defined as $\{a + bi : a, b \in \mathbb{R}\}$ ("$\in$" means "element of"). In other words, complex numbers are an extension of real numbers made by attaching an imaginary number $i$, which satisfies the equality $i^2 = -1$. Complex numbers are of the form $a + bi$, where $a$ and $b$ are real numbers and $i = \sqrt{-1}$. Thus, $a$ is the real part of the number and $b$ is the imaginary part of the number. When $i$ appears in a fraction, the fraction is usually simplified so that $i$ is not in the denominator. The set of complex numbers includes the set of real numbers, where any real number $n$ can be written in its equivalent complex form as $n + 0i$. In other words, it can be said that $\mathbb{R} \subseteq \mathbb{C}$ (or $\mathbb{R}$ is a subset of $\mathbb{C}$).

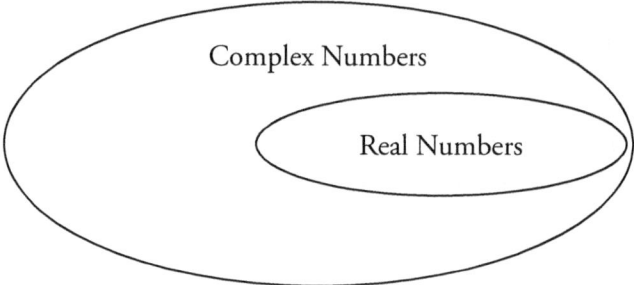

The number $3i$ has a real part 0 and imaginary part 3; the number 4 has a real part 4 and an imaginary part 0. As another way of writing complex numbers, we can express them as ordered pairs:

| Complex number | Ordered pair |
|---|---|
| $3+2i$ | $(3, 2)$ |
| $\sqrt{3} + \sqrt{3}i$ | $(\sqrt{3}, \sqrt{3})$ |
| $7i$ | $(0, 7)$ |
| $\dfrac{6+2i}{7}$ | $\left(\dfrac{6}{7}, \dfrac{2}{7}\right)$ |

The basic operations for complex numbers can be summarized as follows, where $z_1 = a_1 + b_1 i$ and $z_2 = a_2 + b_2 i$. Note that the operations are performed in the standard manner, $i$ being treated as a standard radical value. The result of each operation is written in the standard form for complex numbers. Also note that the complex conjugate of a complex number $z = a + bi$ is denoted as $z^* = a - bi$.

$$z_1 + z_2 = (a_1 + a_2) + (b_1 + b_2)i$$

$$z_1 - z_2 = (a_1 - a_2) + (b_1 - b_2)i$$

$$z_1 z_2 = (a_1 a_2 - b_1 b_2) + (a_1 b_2 + a_2 b_1)i$$

$$\frac{z_1}{z_2} = \frac{z_1 z^*_2}{z_2 z^*_2} = \frac{a_1 a_2 + b_1 b_2}{a_2^2 + b_2^2} + \frac{a_2 b_1 - a_1 b_2}{a_2^2 + b_2^2}i$$

# Chapter 2: Algebra and Functions

## 2.1 Equations and inequalities

Both equations and inequalities relate two quantities which may be expressed as any combination of constants, variables, and functions. In an equation, the two quantities are stated to be equal. In an inequality, one quantity is or may be greater than the other.

### Equations

An **equation** consists of two expressions linked by an equal sign (statement H1) = (statement H2)

Left Hand Side (LHS) = Right Hand Side (RHS).

If substituting a value for the variable results in LHS = RHS, or a true statement, then the value is a solution for that equation.

**Example:**  $2x = 6$

(LHS)  (RHS)

If we substitute 3 for $x$, we get $2 \cdot 3 = 6$ (True).

Therefore, 3 is a solution for the equation.

**Example:**  Is 2 a solution of $2x - 6 = 6x + 1$?

Substituting 2 for $x$, we get

$2(2) - 6 = 6(2) + 1$
$4 - 6 = 12 + 1$
$-2 = 13$ (False)

Therefore, 2 is not a solution.

### Inequalities

An **inequality** has the same form as an equation, but the equals sign is replaced by one of the following inequality signs:

< (less than)
> (greater than)

≤ (less than or equal to)

≥ (greater than or equal to)

The solution to an inequality is not a single value but a set of values that satisfy the inequality.

Example: $x + 2 < 7$

> The solution is $x < 5$, meaning that any number less than 5 is a solution of the inequality.

*Important facts about inequalities*

1. **Sense of an inequality:** This is the direction of the inequality. The larger number is always facing the open side.

    Example: $25 > 3$ (greater than)

    Example: $3 < 25$ (less than)

2. **Notation:**
    $\geq \equiv$ "Greater than *or* equal to".
    $\leq \equiv$ "Less than *or* equal to".
    These relations are satisfied if either half of the relation is satisfied.

    Example: $25 \geq 3$ is true if $25 > 3$ is true or if $25 = 3$ is true. Since $25 > 3$ is true, $25 \geq 3$ is true, even though $25 = 3$ is false.

    Example: $0 \leq 0$ is true if $0 < 0$ is true or $0 = 0$ is true. Since $0 = 0$ is true, $0 \leq 0$ is true, even though $0 < 0$ is false.

3. Multiplying or dividing by a negative number changes the direction of the inequality.

    Example: $-3x > 6$

    > Dividing both sides by $-3$, we get
    >
    > $x < -2$ (note the change in direction)

**Properties of equations and inequalities**

1. We can add any real number to, or subtract any real number from, both sides of the equation (or inequality).

    Example: $3 = 3 \Rightarrow 3 + 2 = 3 + 2 \Rightarrow 5 = 5$ (still true)

    Example: $9 = 9 \Rightarrow 9 - 3 = 9 - 3 \Rightarrow 6 = 6$ (still true)

    Example: $x + 3 = 6 \Rightarrow x + 3 - 3 = 6 - 3 \Rightarrow x = 3$

2. We can multiply or divide both sides of an equation or an inequality by any real number except 0.

*Recall:* When multiplying or dividing by a negative number we change the direction of the inequality.

Example: $3 = 3 \to 3 \times 2 = 3 \times 2 \to 6 = 6$ (still true)

Example: $8 = 8 \to \dfrac{8}{2} = \dfrac{8}{2} \to 4 = 4$ (still true)

Example: $-2x = 6 \to \dfrac{-2x}{-2} = \dfrac{6}{-2} \to x = -3$

Example: $6 > 2 \to 6 \times 2 > 2 \times 2 \to 12 > 4$ (still true)

Example: $-2 < 6 \to \dfrac{-2}{-2} > \dfrac{6}{-2} \to 1 > -3$ (still true, but with reversed inequality)

Example: $-3x \geq 5 \to \dfrac{-3x}{-3} \leq \dfrac{5}{-3} \to x \leq -\dfrac{5}{3}$ (note reversed inequality)

## 2.2 Linear equations

A **linear equation** is one in which no variable has a higher power than 1.

**Solving linear equations**
1. Expand to eliminate all parentheses.
2. If there are fractions, multiply each term by the LCD to eliminate all denominators.
3. Combine terms on each side when possible.
4. Perform operations on both sides of the equation to isolate all variables on one side and all constants on the other side.

Example: Solve for $x$: $3(x + 3) = -2x + 4$

$$
\begin{aligned}
3(x + 9) &= -2x + 4 && \text{Expand parentheses.} \\
3x &= -2x - 5 && \text{Subtract 9 from both sides.} \\
5x &= -5 && \text{Add } 2x \text{ to both sides.} \\
x &= -1 && \text{Divide both sides by 5.}
\end{aligned}
$$

Example: Solve for $x$: $2x + 9 - 3x + 10 = 3x + x - 6$

$$
\begin{aligned}
-x + 19 &= 4x - 6 && \text{Combine similar terms on each side.} \\
-x &= 4x - 25 && \text{Subtract 19 from both sides.} \\
-5x &= -25 && \text{Subtract } 4x \text{ from both sides.} \\
x &= 5 && \text{Divide both sides by } -5.
\end{aligned}
$$

Example: Solve for $x$: $3x - \dfrac{2}{3} = \dfrac{5x}{2} + 2$

$$
\begin{aligned}
18x - 4 &= 15x + 12 && \text{Multiply each term by 6, the LCD of 2 and 3.} \\
18x &= 15x + 16 && \text{Add 4 to each side.} \\
3x &= 16 && \text{Subtract } 15x \text{ from each side.} \\
x &= \dfrac{16}{3} && \text{Divide each side by 3.}
\end{aligned}
$$

## 2.3 Linear inequalities

**Linear inequalities** are inequalities in which no variable has a higher power than one.

### Solving linear inequalities

We use the same procedure used for solving linear equations, but the answer is represented in graphical form on the number line or in interval form.

Example: Solve the inequality, show its solution using interval form, and graph the solution on the number line.

$$\frac{5x}{8} + 3 \geq 2x - 5$$

| | |
|---|---|
| $5x + 24 \geq 16x - 40$ | Multiply each term by 8 to clear denominator. |
| $5x \geq 16x - 64$ | Subtract 24 from each side. |
| $-11x \geq -64$ | Subtract $16x$ from each side. |
| $x \leq 5\frac{9}{11}$ | Divide each side by $-11$; reverse inequality sign. |

Solution in interval form: $(-\infty, 5\frac{9}{11}]$ (Note that "[" means $5\frac{9}{11}$ is included in the solution.)

Graph of solution:

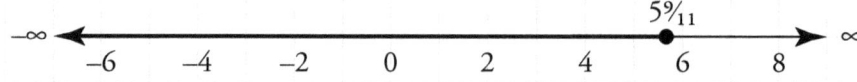

### Interval and graph notation for inequalities

1. [ and ] mean that the lower and upper limit, respectively, are included as solutions. In graphing, a *closed dot* (•) indicates the same thing. Inclusive limits are specified with "greater than or equal to" or "less than or equal to" inequalities.
2. ( and ) mean that the lower and upper limit, respectively, are excluded as solutions. In graphing, *an open* dot (∘) indicates the same thing. Exclusive limits are specified with "greater than" or "less than" inequalities.

Example: Solve the following inequality and express your answer in both interval and graphical form.

| | |
|---|---|
| $3x - 8 < 2(3x - 1)$ | |
| $3x - 8 < 6x - 2$ | Distributive property. |
| $3x < 6x + 6$ | Add 8 to each side. |
| $-3x < 6$ | Subtract $6x$ from each side. |
| $x > -2$ | Divide each side by $-3$; reverse inequality. |

In graphical form:

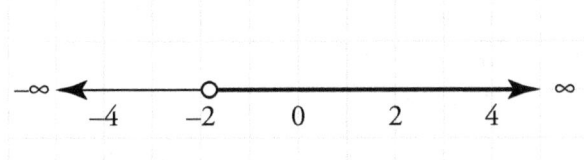

Interval form: (−2,∞) (Note that the "(" means that −2 is NOT included as a solution.)

**Example:** Is −2 one of the solutions of the following inequality?

$2x - 6 \leq x + 4$

Substituting −2 for $x$, we get

$2(-2) - 6 \leq -2 + 4$

$-10 \leq 2$

This is a true statement; therefore, −2 is a solution of the inequality.

**Example:** Is 3 one of the solutions of the following inequality?

$3x \leq 3 + 2$

Substituting 3 for $x$, we get

$3(3) \leq 3 + 2$

$9 \leq 5$

This statement is false; therefore, 3 is not a solution of the inequality.

**Note:**  a. A linear equation has one solution, no solution, or an infinite number of solutions.
b. A linear inequality can have any number of solutions.

## Graphing a linear inequality

A linear inequality in two variables is similar in form to a linear equation *except* that the = sign is replaced by an inequality sign of >, <, ≥, or ≤. The procedure to graph it is as follows:

1. Graph the equivalent equation with the inequality sign replaced by an equals sign. Use a solid line for this line if the inequality contains the equals sign (≤ or ≥); use a dashed line if the inequality contains no equals sign (< or >).
2. Pick a point on either side of the line and test whether its $x$- and $y$-values satisfy the inequality. If so, mark that region as a solution set with shading or slanted lines. If not, shade the opposite region.

**Example:** Identify the region that satisfies $3x + 5y < 15$.

1. We graph the equivalent equation of $3x + 5y = 15$ using a dashed line. Substituting $y = 0$ produces an $x$-intercept of $(5, 0)$; substituting $x = 0$ produces a $y$-intercept of $(0,3)$.

2. Pick a test point on either side of this line. Pick the origin for simplicity $(0, 0)$. Substitute $x = 0$ and $y = 0$ into the inequality and test:

$0 < 15$ is true, so accept the region containing $(0, 0)$ and shade it.

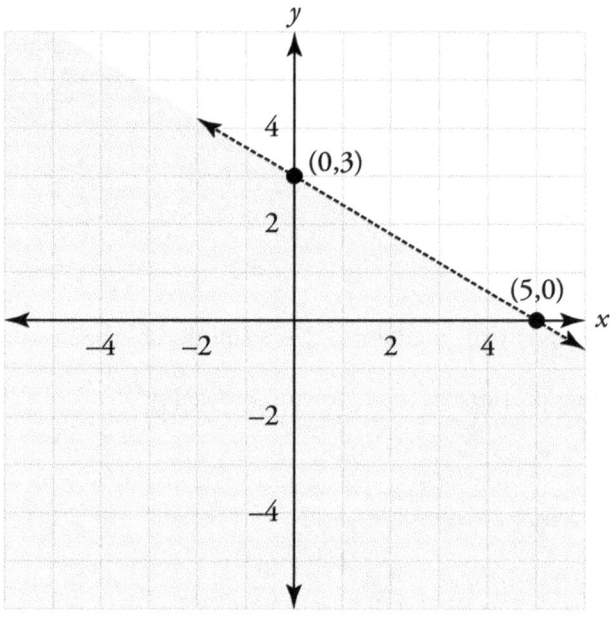

### Use of "and" and "or" with inequalities

1. "And" means "intersection," the overlapping of two regions. It indicates the region that is common to the solutions of more than one inequality.
2. "Or" means "union," the joining of two regions. It indicates the region that is a solution of either inequality or both.

**Example:** Solve the inequalities $3x = 5y < 15$ and $y \geq 1$.

1. We found the region representing the inequality $3x + 5y < 15$ in example 2. Shade it with horizontal lines.
2. For $y \geq 1$, graph the equivalent equation of $y = 1$. This is a horizontal line through $y = 1$. Use a solid line.
3. Use the origin $(0, 0)$ as a test point. Test the inequality with $x = 0$, $y = 0$: $0 \geq 1$ is false. So shade the region above the line with vertical lines.
4. The solution region is the intersection region where the two regions overlap as shown.

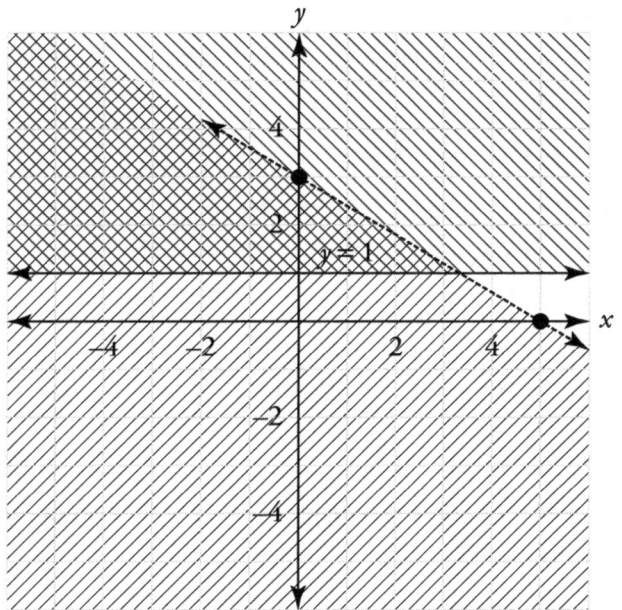

If the question had called for the solution of $3x + 5y < 15$ OR $y \geq 1$, the solution region would be as shown below.

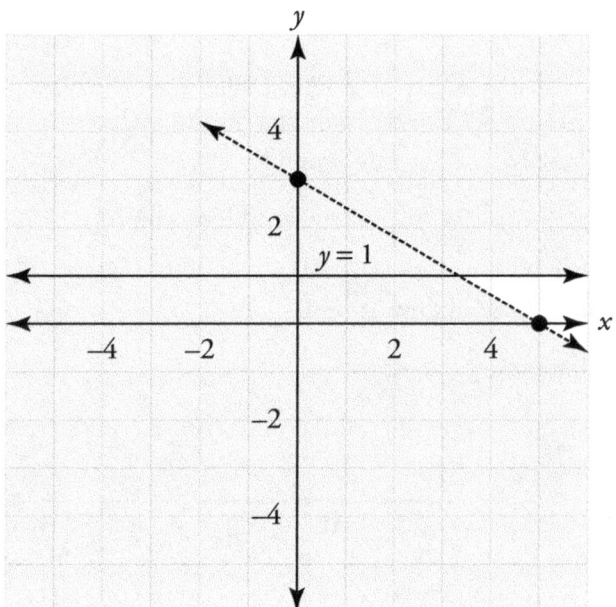

## 2.4 Absolute-value equations and inequalities

An **absolute-value equation** or **inequality** is one in which one or more of the variables is given as an absolute value. Solving absolute-value equations and inequalities introduces a new level of complexity, because in many cases it is necessary to consider two different values of an expression whose absolute value would be the same and to solve separately for each one.

## Absolute-value equations

If $a$ and $b$ are real numbers, and $k$ is a nonnegative real number, the solution of $|ax + b| = k$ is $ax + b = k$ or $ax + b = -k$

**Example:** Solve for $x$: $|2x + 3| = 9$

$$2x + 3 = 9 \qquad \text{or} \qquad 2x + 3 = -9$$
$$2x + 3 - 3 = 9 - 3 \qquad 2x + 3 - 3 = -9 - 3$$
$$2x = 6 \qquad\qquad 2x = -12$$
$$x = 3 \qquad\qquad x = -6$$

Therefore, the solution is $x = \{3, -6\}$

**Example:** Solve for $x$: $|3x - 1| = -3$

Since $-3$ is a negative number, and an absolute value cannot be negative, there is no solution.

## Absolute-value inequalities

If $a$ and $b$ are real numbers and $k$ is a nonnegative real number, the solution of $|ax + b| < k$ is $-k < ax + b < k$

**Example:** Solve $|7x + 3| < 25$

$$-25 < (7x + 3) < 25 \qquad \text{Rewrite original inequality.}$$
$$(-25 - 3) < 7x < (25 - 3) \qquad \text{Subtract 3 from each term.}$$
$$-28 < 7x < 22 \qquad \text{Simplify.}$$
$$-4 < x < \frac{22}{7} \qquad \text{Divide all terms by 7.}$$

Solution in interval form is $\left(-4, \frac{22}{7}\right)$.

In graphic form:

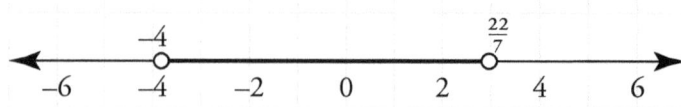

If $a$ and $b$ are real numbers and $k$ is a nonnegative real number, the solution of $|ax + b| > k$ is $ax + b > k$ or $ax + b < -k$

**Example:** Solve $|2x - 7| > 5$

$$2x - 7 > 5 \qquad\qquad 2x - 7 < -5$$
$$2x - 7 + 7 > 5 + 7 \qquad 2x - 7 + 7 < -5 + 7$$
$$2x > 12 \qquad\qquad 2x = 2$$
$$x > 6 \qquad\qquad x < 1$$

Solution: $x > 6$ or $x < 1$

In interval form: $(-\infty, 1) \cup (6, \infty)$

Graphically:

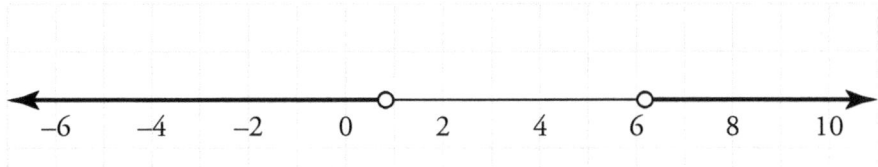

## 2.5 Functions

An equation like $y = 3x + 5$ describes a relation between the independent variable $x$ and the dependent variable $y$. Thus $y$ is written as $f(x)$, "function of $x$." But $y$ is only a true **function** if there is a relationship between the set of all inputs or values of the independent variable (the domain) and the set of all outputs or values of the dependent variable (the range) such that each element of the domain corresponds to one element of the range. (For any input there is exactly one output.)

**Example:**

| $x$ | $y$ |
|---|---|
| 2 | 4 |
| 4 | 8 |
| 8 | 16 |
| (This is a function.) | |

| $x$ | $y$ |
|---|---|
| 3 | 7 |
| 3 | 10 |
| 6 | 13 |
| (This is NOT a function.) | |

**Example:** Given the function $f(x) = 3x + 5$:

Find $f(2); f(0); f(-10)$

Finding $f(2)$ means finding the function value at $x = 2$.

For $f(x) = 3x + 5$:

$f(2) = 3(2) + 5 = 6 + 5 = 11$

$f(0) = 3(0) + 5 = 0 + 5 = 5$

$f(-10) = 3(-10) + 5 = -30 + 5 = -25$

### Properties of functions

A **relation** is any set of ordered pairs. The domain of a relation is the set containing all the first coordinates of the ordered pairs, and the range of a relation is the set containing all the second coordinates of the ordered pairs.

A function is a relation in which each value in the domain corresponds to only one value in the range. It is notable, however, that a value in the range may correspond to any number of values in the domain. Thus, although a function is necessarily a relation, not all relations are functions, since a relation is not bound by this rule.

On a graph, use the vertical line test to check whether a relation is a function. If any vertical line intersects the graph of a relation in more than one point, as in the graph below, then the relation is not a function.

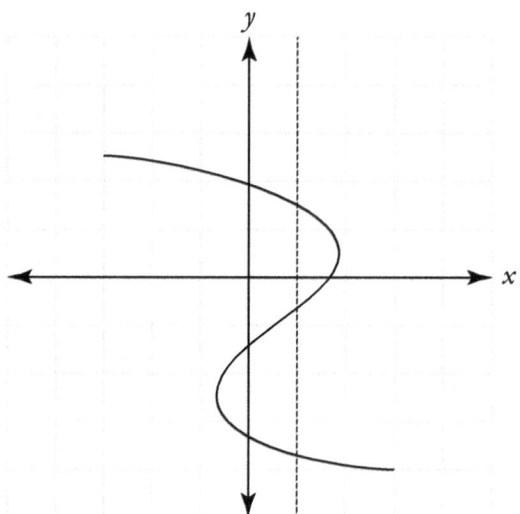

A relation is considered one-to-one if each value in the domain corresponds to only one value in the range, and each value in the range corresponds to only one value in the domain. Thus, a one-to-one relation is also a function, but it adds an additional condition.

In the same way that the graph of a relation can be examined using the vertical line test to determine whether it is a function, the horizontal line test can be used to determine if a function is a one-to-one relation. If no horizontal lines superimposed on the plot intersect the graph of the relation in more than one place, then the relation is one-to-one (assuming it also passes the vertical line test and, therefore, is a function).

As mentioned above, a function is a relation in which each value in the domain corresponds to only one value in the range. Functions can be expressed discretely, as sets of ordered pairs, or they can be expressed more generally as formulas. For instance, the function $y = x$ is a function that represents an infinite set of ordered pairs $(x, y)$, where each value in the domain ($x$) corresponds to the same value in the range ($y$).

### Families of functions

Some of the most commonly used function families include linear, polynomial, rational, exponential, logarithmic, and trigonometric functions. These functions, separately or in various combinations, can be used to model a range of common phenomena in finance, physics, and other fields.

### Linear functions

A **linear function** can be expressed as $f(x) = mx + b$, where $m$ and $b$ are constants. It is called linear because it involves no quadratic or cubic variables, nor any square roots or cube roots of variables. No variables in a linear function have any exponent other than 1.

A linear function can be graphed as $y = mx + b$. The result is a straight line with slope $m$ that intercepts the $y$-axis at $(0, b)$.

**Example:** $y = 2x - 1$

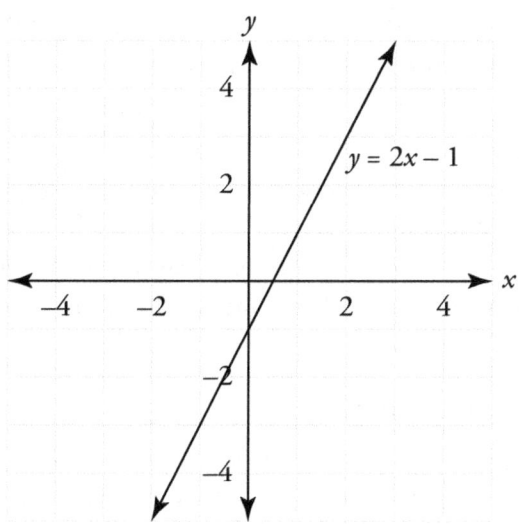

## Direct variation

If, in a function $y = mx + b$, $b$ is 0, the relation is a **direct variation**. In a direct variation, $y$ and $x$ are always in the same proportion. That means that there is a constant $c$ such that $y = cx$. Theoretically, $c$ could be positive or negative, but in actual practice, $c$ is almost always positive, which means that as one parameter gets larger, so does the other. The graph of a direct variation always passes through $(0, 0)$.

**Example:** A brand of ketchup contains 3g of sugar per ounce. This is a direct variation. If the ounces of ketchup are plotted as $x$ and the grams of sugar as $y$, then $y = 3x$ as in the graph below.

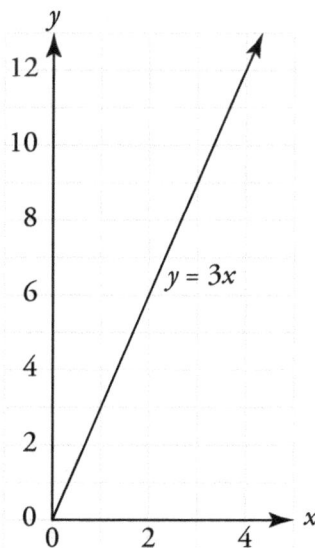

### Domain and range

The **domain** of a function is the set of all possible inputs to the function. The **range** of a function is the set of all possible outputs. In some functions, both the domain and the range extend to all real numbers. Some functions have limitations on the domain, meaning that certain values are not allowed as inputs. Some functions have limitations on the range, meaning that certain values are not possible as outputs.

In the function $y = 2x + 4$, both the domain and the range extend to all real numbers. Any real number is a possible value of the input $x$ or the output $y$.

In the function $y = \frac{1}{x-3}$, the domain includes all real numbers except 3: $x$ cannot equal 3, because that would cause a division by zero.

In the function $y = x^2$, the range is all positive real numbers. Since the square of a real number is always positive, the possible outputs of the function do not include any negative numbers.

## 2.6 Systems of linear equations

A **system of equations** is a group of equations which are simultaneously true. These equations may involve one, two or more variables. A system of equations that has the same number of equations as there are variables is called a "square system". A square system will have unique solutions for each variable. A nonsquare system may have no solution at all or infinitely many solutions. We will explore three basic ways of solving systems of linear equations: graphing, elimination, and substitution.

### Solving a system of equations by elimination

Here we use a method which results in the elimination of one variable; this reduces the system to one equation.

**Example:** Solve by the elimination method.

$$\begin{cases} 3x + 5y = 10 & [1] \\ 2x + 3y = 7 & [2] \end{cases}$$

Multiply $3x$ by 2 to get $6x$, multiply $2x$ by $-3$ to get $-6x$.

Add the new equations to eliminate $x$.

Eq. 1 × 2 $\Rightarrow$ $6x + 10y = 20$     [3]

Eq. 2 × −3 $\Rightarrow$ $-6x - 9y = -21$     [4]

Adding [3] and [4], we get $y = -1$ [5]

Substitute [5] into [1] and solve for $x$

$$3x + 5(-1) = 10$$
$$3x - 5 = 10$$
$$3x = 15$$
$$x = 5$$

Check solutions using equation [2]:

$2(5) + 3(-1) = 7$

$10 - 3 = 7$ ✓

Solution is $x = 5, y = -1$.

The answer can be represented by $(5, -1)$, since it represents the point of intersection of the two lines.

## Solving a system of equations by substitution

Here we rewrite one equation in terms of a single variable. Then we substitute the expression of the variable into the second equation.

Example: Solve the system by the substitution method:

$$\begin{cases} x - y = 1 & [1] \\ 3x + 2y = 38 & [2] \end{cases}$$

from [1], adding $y$ to each side produces

$x = y + 1$      [3]

substitute [3] into [2]

$3(y+1) + 2y = 38$
$3y + 3 + 2y = 38$
$5y + 3 = 38$
$5y = 35$
$y = 7$      [4]

Substituting [4] into 3] produces

$x = 7 + 1 = 8$

The solution is $x = 8, y = 7$ or $(8,7)$.

Note: When solving a linear system of equations the elimination method is easier to use for most if not all problems.

## Systems of equations with infinitely many solutions

If a system of equations reduces to a true statement, such as $3 = 3$, the system has infinitely many solutions.

Example: Solve the system

$$\begin{cases} 2m + 5n = 1 & [1] \\ 6m + 15n = 3 & [2] \end{cases}$$

Multiply [1] by −3: $-6m - 15n = -3$    [3]

Adding [3] and [2] gives

$0 = 0$, which is true for any values of $x$. Therefore, there are infinitely many solutions. The two equations represent the same line.

### Systems of equations with no solution

**Example:** Solve the system

$$\begin{cases} 7x + 5y = 25 & [1] \\ 14x + 10y = -30 & [2] \end{cases}$$

Multiplying [1] by −2 gives

$-14x - 10y = -50$    [3]

Adding [2] and [3] gives

$0 = -80$, which is false.

Therefore, there is no solution. The two lines are parallel and do not intersect.

## 2.7  Systems of linear inequalities

Solving systems of linear inequalities is best performed graphically. To graph a linear inequality expressed in terms of $x$ and $y$, solve the inequality for $y$. This renders the inequality in slope-intercept form ($y = mx + b$). The point $(0, b)$ is the $y$-intercept, and $m$ is the slope of the line. If the inequality is expressed only in terms of $x$, solve for $x$. When solving an inequality, remember that dividing or multiplying by a negative number will reverse the direction of the inequality sign.

If an inequality yields any of the following results in terms of $y$, where $a$ is some real number, the solution set of the inequality is bounded by a *horizontal line*:

$y < a, y \leq a, y > a, y \geq a$

If the inequality yields any of the following results in terms of $x$, then the solution set of the inequality is bounded by a *vertical line*:

$x < a, x \leq a, x > a, x \geq a$

When graphing the solution of a linear inequality, the boundary is drawn as a dashed line if the inequality sign is < or >. This indicates that points on the line do not satisfy the inequality. If the inequality sign is either ≤ or ≥, then the boundary is drawn as a solid line to indicate that the points on the line satisfy the inequality.

The line drawn as directed above is only the boundary of the solution set for an inequality. The solutions actually include the half plane bounded by the line.

Since, for any line, half of the values in the full plane (for either $x$ or $y$) are greater than those defined by the line and half are less, the solution of the inequality always amounts to a half plane. Which half plane satisfies the inequality can be found by testing a point on either side of the line. The solution set can be indicated on a graph by shading the appropriate half plane.

For inequalities expressed as a function of *x*, shade above the line when the inequality sign is > or ≥. Shade below the line when the inequality sign is < or ≤.

For inequalities expressed as a function of *y*, shade to the right for > or ≥. Shade to the left for < or ≤.

The solution to a system of linear inequalities consists of the portion of the graph where the shaded half planes for all the inequalities in the system overlap. For instance, if the graph of one inequality was shaded with red and the graph of another inequality was shaded with blue, then the overlapping area would be shaded purple. The points in the purple area would be the solution set of this system.

**Example:** Solve by graphing:

$$\begin{cases} x + y \leq 6 \\ x - 2y \leq 6 \end{cases}$$

Solving the inequalities for *y*, they become

$$\begin{cases} y \leq -x + 6 \text{ (slope } -1, y\text{-intercept } 6) \\ y \geq \frac{1}{2}x - 3 \text{ (slope } \frac{1}{2}, y\text{-intercept } -3) \end{cases}$$

A graph with the appropriate shading is shown below:

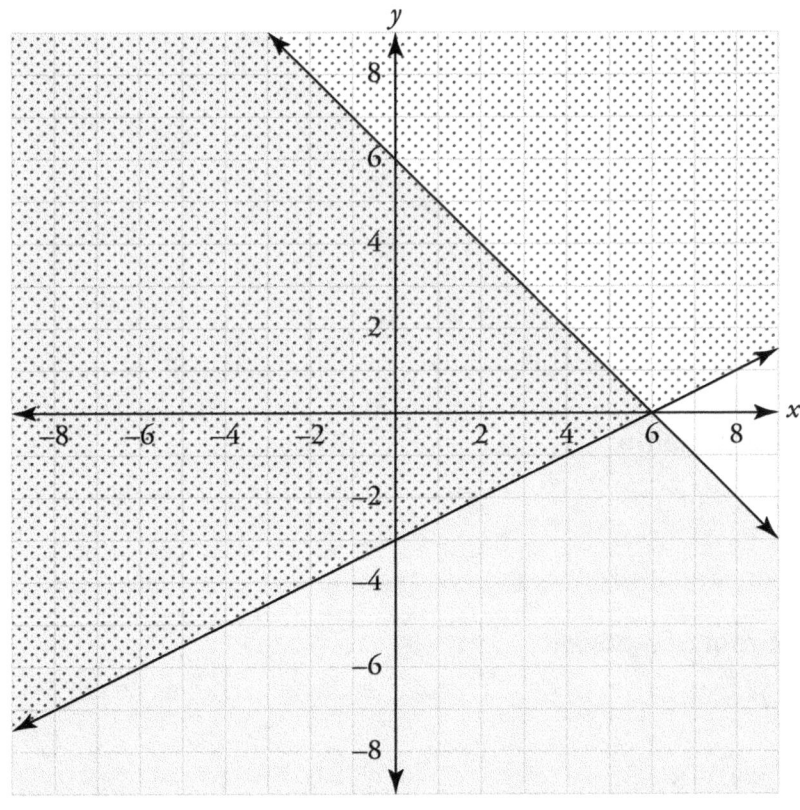

Chapter 2: Algebra and Functions

## 2.8 Operations with exponents

The **exponent** of a number indicates how many factors of that number are being multiplied together: $5^2 = 5 \times 5$, $5^3 = 5 \times 5 \times 5$ and so on. The following operations apply to numbers with exponents.

### Multiplying two powers
$$a^m \times a^n = a^{m+n}$$

**Example:** $3^3 \times 3^2 = 3^{3+2} = 3^5 = 243$

### Dividing two powers
$$\frac{a^m}{a^n} = a^{m-n}$$

**Example:** $\frac{2^8}{2^5} = 2^{8-5} = 2^3 = 8$

### Raising a power to a power
$$(a^m)^n = a^{mn}$$

**Example:** $(3^2)^3 = 3^{2 \cdot 3} = 3^6 = 729$

### Raising a product to a power
$$(ab)^m = a^m b^m$$

**Example:** $(2 \cdot 3)^3 = (2^3)(3^3) = 216$

### Negative exponents
$$a^{-n} = \frac{1}{a^n} = \left(\frac{1}{a}\right)^n$$

**Example:** $2^{-3} = \frac{1}{2^3} = \frac{1}{8}$

### Zero exponents
If $a$ is any nonzero number then $a^0 = 1$.

**Example:** $257^0 = 1$

### Fractional exponents
$$a^{\frac{m}{n}} = (a^{\frac{1}{n}})^m = (a^m)^{\frac{1}{n}}$$

**Example:** $9^{\frac{3}{2}} = (9^{\frac{1}{2}})^3 = (9^3)^{\frac{1}{2}} = 27$

### Taking the root of a power
$$\sqrt[n]{a^m} = a^{\frac{m}{n}}$$

**Example:** $\sqrt[3]{8^2} = 8^{\frac{2}{3}} = 4$

## 2.9 Properties of logarithms

**Exponential** and **logarithmic** functions are complementary. The general relationship for logarithmic and exponential functions is as follows.

$y = \log_b x$ if and only if $x = b^y$

**Example:** $125 = 5^3$, therefore $\log_5(125) = 3$

The relationship is as follows for the exponential base $e$ and the natural logarithm (ln).

$y = \ln x$ if and only if $e^y = x$

**Example:** $e^{3.5835} \approx 36$, therefore $\ln 36 \approx 3.5835$

The following properties of logarithms are helpful in solving equations.

### Multiplication property

$\log_b mn = \log_b m + \log_b n$

**Example:** $\log_4(2) = 0.5$, $\log_4(8) = 1.5$, $\log_4(2 \cdot 8) = 0.5 + 1.5 = 2$

### Quotient property

$\log_b \dfrac{m}{n} = \log_b m - \log_b n$

**Example:** $\log_9(243) = 2.5, \log_9(3) = 0.5, \log_9\left(\dfrac{243}{3}\right) = 2.5 - 0.5 = 2$

### Powers property

$\log_b n^r = r \log_b n$

**Example:** $\log_2(4) = 2$, $\log_2(4^3) = 3(2) = 6$

### Equality property

$\log_b n = \log_b m$ if and only if $n = m$

**Example:** if $\log_{16}(m) = \log_{16}(64)$, $m = 64$

### Change of base formula

$\log_b n = \dfrac{\log_a n}{\log_a b}$

**Example:** $\log_3(25) = \dfrac{\log_{10}(25)}{\log_{10}(3)}$

$\log_b b^x = x$

**Example:** $\log_6(6^3) = 3$

$b^{\log_b x} = x$

**Example:** $6^{\log_6(216)} = 216$

Chapter 2: Algebra and Functions

## 2.10 Solving problems involving exponential or logarithmic functions

Solving problems involving exponentials or logarithms typically involves isolating the terms containing the exponential or logarithmic function and using the inverse operation to "extract" the argument. For instance, given the following equation,

$\ln f(x) = c$

the function $f(x)$ can be determined by raising $e$ to each side of the equation.

$e^{\ln f(x)} = f(x) = e^c$

Alternatively, if the function is in terms of an exponent $e$,

$e^{f(x)} = c$

solve by taking the natural logarithm of both sides.

$\ln e^{f(x)} = f(x) = \ln c$

Although these examples are in terms of $e$ and natural logarithms, the same logic applies to exponentials and logarithms involving different bases as well.

## 2.11 Expanding polynomials

To multiply two binomials, use the FOIL method: First terms, Outside terms, Inside terms, Last terms.

**Example:** expand $(3x + 1)(x - 7)$.

$$(3x + 1)(x - 7) =$$
$$\text{First terms: } 3x(x) +$$
$$\text{Outside terms: } 3x(-7) +$$
$$\text{Inside terms: } 1(x) +$$
$$\text{Last terms: } 1(-7)$$
$$= 3x^2 + (-21x) + x - 7$$
$$= 3x^2 - 20x - 7$$

## 2.12 Factoring polynomials

Polynomials, like integers, can be factored. The factors of a polynomial can include constants, variables, and expressions. There are various strategies for factoring a polynomial, depending on the form it's in.

### Removing common factors

The first step in factoring any polynomial is to remove any common factor of all terms, using the distributive property.

**Example:** $8x^3 - 18x^2 - 4x = 2x(4x^2 - 9x - 2)$

### Factoring the difference of two squares

If a polynomial has two terms, check if it is the difference of two squares. If it can be expressed as $a^2 - b^2$, it can be factored as $(a + b)(a - b)$.

Example:  $9x^2 - 4 = (3x)^2 - (2)^2 = (3x + 2)(3x - 2)$

## Factoring a trinomial in the form $x^2 + bx + c$

Look for two factors of $c$ that add to $b$.

Example:  factor $x^2 + 11x + 18$

        Factors of 18: 1 and 18 (sum 19) ✗

                    2 and 9 (sum 11) ✓

        $x^2 + 11x + 18 = (x + 2)(x + 9)$

## Factoring a trinomial in the form $x^2 - bx + c$

Look for two factors of $c$ that add to the absolute value of $b$, then make both factors negative.

Example:  $x^2 - 11x + 18 = (x - 2)(x - 9)$

## Factoring a trinomial in the form $x^2 + bx - c$

Look for two factors of the absolute value of $c$ that differ by $b$. Make the larger factor positive and the smaller factor negative.

Example:  factor $x^2 + 5x - 24$

        Factors of 24: 1 and 24 (difference 23) ✗

                    2 and 12 (difference 10) ✗

                    3 and 8 (difference 5) ✓

        $x^2 + 5x - 24 = (x + 8)(x - 3)$

## Factoring a trinomial in the form $x^2 - bx - c$

Look for two factors of the absolute value of $c$ that differ by the absolute value of $b$. Make the larger factor negative and the smaller factor positive.

Example:  $x^2 - 5x - 24 = (x - 8)(x + 3)$

## Factoring a trinomial in the form $ax^2 + bx + c$

Look for two factors of $ac$ that add to $b$.

Example:  factor $4x^2 + 8x + 3$

        Factors of 12: 1 and 12 (sum 13) ✗

                    2 and 6 (sum 8) ✓

$4x^2 + 8x + 3 = 4x^2 + 2x + 6x + 3$    Rewrite the $x$-term as two terms, using the factors you have found.

$= 2x(2x + 1) + 6x + 3)$    Factor out the largest common factor of the first two terms.

$= 2x(2x + 1) + 3(2x + 1)$    Factor out the binomial in parentheses from the last two terms.

$= (2x + 3)(2x + 1)$    Use the distributive property to gather together the two terms of the other binomial factor.

### Factoring a trinomial in the form $ax^2 - bx + c$

**Example:** factor $4x^2 - 8x + 3$

Look for two factors of $ac$ that add to the absolute value of $b$, make them both negative, and rewrite the polynomial, separating the $x$-term into two factors corresponding to the two negative factors.

$4x^2 - 8x + 3 = 4x^2 - 6x - 2x + 3$    Rewrite the $x$-term as two terms, using the factors you have found.

$= 2x(2x - 3) - 2x + 3)$    Remove the largest common factor from the first two terms.

$= 2x(2x - 3) - 1(2x - 3)$    Factor out the binomial in parentheses from the last two terms.

$= (2x - 1)(2x - 3)$    Use the distributive property to gather together the two terms of the other binomial factor.

### Factoring a trinomial in the form $ax^2 + bx - c$

**Example:** factor $9x^2 + 3x - 2$

Look for two factors of the absolute value of $ac$ that differ by $b$.

Factors of 18: 1 and 18 (difference 17) ✗

2 and 9 (difference 7) ✗

3 and 6 (difference 3) ✓

Make the larger factor positive and the smaller factor negative. Rewrite the polynomial, separating the $x$-term into terms corresponding to the two factors, positive and negative.

$$9x^2 + 3x - 2 = 9x^2 + 6x - 3x - 2$$ Rewrite the *x*-term as two terms, using the factors you have found.
$$= 3x(3x + 2) - 3x - 2)$$ Factor out the largest common factor of the first two terms.
$$= 3x(3x + 2) - 1(3x + 2)$$ Factor out the binomial in parentheses from the last two terms.
$$= (3x - 1)(3x + 2)$$ Use the distributive property to gather together the two terms of the other binomial factor.

### Factoring a trinomial in the form $ax^2 - bx - c$

Look for two factors of the absolute value of $ac$ that differ by the absolute value of $b$. Make the larger factor negative and the smaller factor positive. Rewrite the polynomial, separating the *x*-term into terms corresponding to the two factors, positive and negative.

**Example:** factor $9x^2 - 3x - 2$

$$9x^2 - 3x - 2 = 9x^2 - 6x + 3x - 2$$ Rewrite the *x*-term as two terms, using the factors you have found.
$$= 3x(3x - 2) + 3x - 2)$$ Factor out the largest common factor of the first two terms.
$$= 3x(3x - 2) + 1(3x - 2)$$ Factor out the binomial in parentheses from the last two terms.
$$= (3x + 1)(3x - 2)$$ Use the distributive property to gather together the two terms of the other binomial factor.

### Factoring polynomials with four terms

Polynomials with four terms can be factored with a technique very similar to that used for trinomials:

**Example:** factor $8x^3 + 12x^2 - 10x - 15$

$$8x^3 + 12x^2 - 10x - 15 = 4x(2x + 3) - 10x - 15$$ Factor out the largest common factor of the first two terms.
$$= 4x^2(2x + 3) - 5(2x + 3)$$ Factor out the binomial in parentheses from the last two terms.
$$= (4x^2 - 5)(2x + 3)$$ Use the distributive property to gather together the two terms of the other binomial factor.

## 2.13 Quadratic equations and quadratic expressions

A **quadratic equation** is one that includes at least one squared term such as $2x^2$.

***Definition:*** The standard form of a quadratic equation is represented by $ax^2 + bx + c = 0$, where $a$, $b$, and $c$ are real, imaginary, or complex numbers and $a \neq 0$.

*Note:* Real numbers are a subset of complex numbers.

Examples: $3x^2 + 5x - 7 = 0$

$-3x^2 + 2x = 0$

$x^2 - 2 = 0$

A **quadratic expression** is one that includes at least one squared term; it is equivalent to the left-hand side of a quadratic equation.

Examples: $2x^2 - 2x + 9$

$4x^2 - 3x$

$-2x^2 + 6$

Quadratic equations may be solved by several methods, including taking the square root, factoring, and using the quadratic formula.

### Taking the square root

If it is possible to take the square root of both sides, a quadratic equation may be solved that way.

Example: $4x^2 + 4x + 1 = 49$

| | | |
|---|---|---|
| $(2x + 1)^2 = 7^2$ | | Represent both sides as squares. |
| $2x + 1 = \pm 7$ | | Take square root of each side. |
| $2x + 1 = 7$ | $2x + 1 = -7$ | Find two solutions. |
| $2x = 6$ | $2x = -8$ | Subtract 1 from each side. |
| $x = 3$ | $x = -4$ | Divide each side by 2. |
| $x = 3, -4$ | | Combine solutions. |

### Factoring

If the left-hand side of a quadratic equation in standard form can be factored, the equation can be solved by setting each factor to 0.

Example: solve for $x$: $x^2 - x - 6 = 0$

| | | |
|---|---|---|
| $(x + 2)(x - 3) = 0$ | | Equation is true if either factor equals 0. |
| $x + 2 = 0$ | $x - 3 = 0$ | Solve by setting each factor equal to 0. |
| $x = -2$ | $x = 3$ | Find two solutions. |
| $x = -2, 3$ | | Combine solutions. |

In general, if $x = a$ is a solution of a quadratic equation in standard form, then $x - a$ is a factor of the expression on the left-hand side of the equation, and vice versa.

Example: 3 is a solution of $x^2 + 2x - 15 = 0$, since $3^2 + 2(3) - 15 = 9 + 6 - 15 = 0$.
Therefore, $(x - 3)$ is a factor of $x^2 + 2x - 15$: $x^2 + 2x - 15 = (x - 3)(x + 5)$

## Using the quadratic formula

When a quadratic equation is not factorable we use the quadratic formula, which *always yields* a solution. The solution of $ax^2 + bx + c = 0$ is given by the formula:

$$x = \frac{-b \pm \sqrt{b^2 - 4ac}}{2a}$$

This formula yields two solutions:

$$x_1 = \frac{-b + \sqrt{b^2 - 4ac}}{2a}, \quad x_2 = \frac{-b - \sqrt{b^2 - 4ac}}{2a}$$

Example: Solve for $x$: $3x^2 + 5x - 3 = 0$

$a = 3; b = 5; c = -3$

$$x = \frac{-b \pm \sqrt{b^2 - 4ac}}{2a}$$

$$= \frac{-5 \pm \sqrt{5^2 - 4(3)(-3)}}{2(3)}$$

$$= \frac{-5 \pm \sqrt{61}}{6}$$

$$x_1 = \frac{-5 + \sqrt{61}}{6}, \quad x_2 = \frac{-5 - \sqrt{61}}{6}$$

## 2.14 Modeling nonlinear functions from real-world data

Various real-world situations provide datasets that can be modeled using various types of functions. In some cases, the situation dictates the sort of function that would model it correctly. For instance, if you shoot a rocket into the air, the data showing its height as a function of time would form a parabola. A growth in volume might be modeled by a cubic function. Populations of animals or plants may expand exponentially for a time if they do not face adverse factors. Compound interest as a function of time is inherently an exponential function. A function that rises and falls, such as the height of ocean tides, might be modeled by a trigonometric function.

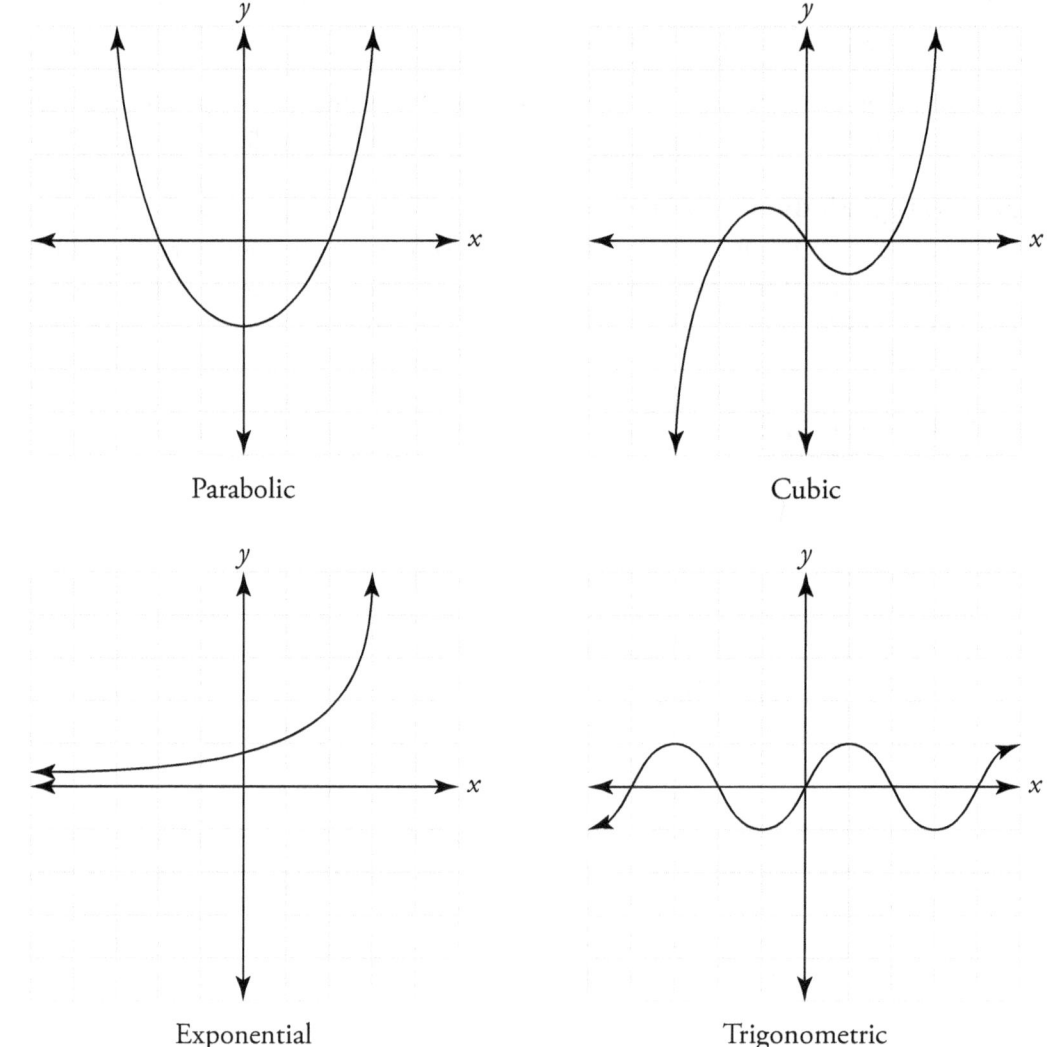

Parabolic

Cubic

Exponential

Trigonometric

In other cases, it may take some investigation and trial-and-error to find what kind of function best models a particular dataset. Sales data for a new product, for instance, might seem exponential at first and then level off as the market becomes more saturated. The number of visits to a particular website, for another example, may be influenced by numerous unseen factors that cannot be quantified separately.

In cases where the nature of the data does not clearly suggest a type of modeling function, a graphing calculator is an enormous aid. With a graphing calculator, you can enter your data points as you find them, and then try various modeling functions for the best possible fit, including linear, quadratic, cubic, and exponential and trigonometric models.

## 2.15  Recursive patterns and relations

A **recurrence relation** is an equation that defines a sequence **recursively**; in other words, each term of the sequence is defined as a function of the preceding terms. For instance, the formula for the balance of an interest-bearing savings account after $t$ years, which is given in a later section in closed form (that is, explicit form), can be expressed recursively as follows.

$$A_i = A_{i-1}\left(1 + \frac{r}{n}\right)^n, \text{ where } A_0 = P, \text{ the initial principal invested.}$$

Here $r$ is the annual interest rate and $n$ is the number of times the interest is compounded per year. Mortgage and annuity parameters can also be expressed in recursive form. Calculation of a past or future term by applying a recursive formula multiple times is called iteration.

Sequences of numbers can be defined by iteratively applying a recursive pattern.

For instance, the **Fibonacci sequence** is defined as follows.

$$F_i = F_{i-1} + F_{i-2} \text{ where } F_0 = 0 \text{ and } F_1 = 1$$

Applying this recursive formula gives the sequence {0, 1, 1, 2, 3, 5, 8, 13, 21, …}. It is sometimes difficult or impossible to write recursive relations in explicit or closed form. In such cases, especially when computer programming is involved, the recursive form can still be helpful. When the elements of a sequence of numbers or values depend on one or more previous values, then it is possible that a recursive formula could be used to summarize the sequence.

If a value or number from a later point in the sequence (that is, other than the beginning) is known and it is necessary to find previous terms, then the indices of the recursive relation can be adjusted to find previous values instead of later ones.

Consider, for instance, the Fibonacci sequence.

$$F_i = F_{i-1} + F_{i-2}$$
$$F_{i+2} = F_{i+1} + F_i$$
$$F_i = F_{i+2} - F_{i+1}$$

Thus, if any two consecutive numbers in the Fibonacci sequence are known, then the previous numbers of the sequence can be found (in addition to the later numbers).

**Example:** Write a recursive formula for the following sequence: {2, 3, 5, 9, 17, 33, 65, …}.

By inspection, it can be seen that each number in the sequence is equal to twice the previous number, less 1. If the numbers in the sequence are indexed such that, for the first number, $I = 1$, and so on, then the recursion relation is the following.

$$N_i = 2N_{i-1} - 1$$

**Example:** If a recursive relation is defined by $N_i = N_{i-1}^2$, and the fourth term is 65,536, what is the first term?

Adjust the indices of the recursion and then solve for $N_i$.

$$N_{i+1} = N_i^2$$
$$N_i = \sqrt{N_{i+1}}$$

Use this relationship to backtrack to the first term.

Chapter 2: Algebra and Functions

$$N_3 = \sqrt{N_4} = \sqrt{65{,}536} = 256$$
$$N_2 = \sqrt{N_3} = \sqrt{256} = 16$$
$$N_1 = \sqrt{N_2} = \sqrt{16} = 4$$

The first term of the sequence is 4.

## 2.16 Parametric functions

In the most familiar kind of function, $x$ is the input and $y$ is the output. For instance, if $y = x^2 + 1$, then an input of $x = 3$ produces an output of $y = 10$.

In a **parametric function**, the input is a separate variable (or parameter), typically represented by $t$, and $x$ and $y$ are both derived from $t$ using separate equations, which are called parametric equations.

For example, suppose you collect model cars. Each car costs $20 and requires 1.5 ft² of storage space. If $x$ is the cost in dollars and $y$ is the square feet of space required, both can be related to $t$, the number of cars you have at any given time.

$x = 20t$

$y = 1.5t$

Parametric functions are a good way to model real-world situations in which two factors are controlled by a third. In this case, a function directly linking $y$ to $x$ would not really describe the situation. The space required does not increase because more money is spent. Rather, both the space required and the money spent depend on the number of cars. Modeling the situation as a parametric function makes that clear.

A function in the usual form can be converted to a pair of parametric equations by setting $x$ equal to $t$ and replacing $x$ with $t$ in the original function.

**Example:** Convert $y = x^2 + 4x + 4$ to parametric form.

$x = t$

$y = t^2 + 4t + 4$

By using $t$ to represent time, parametric functions can be used to model movements that are at different speeds as well as in different directions.

**Example:** Two ants are on a piece of graph paper. One ant is at $(-2, -3)$. Its motion can be described by the following parametric equations:

$x_1 = 2t - 2$

$y_1 = 4t - 3$

The second ant is at $(-4, -1)$. Its motion can be described as follows:

$x_2 = t - 4$

$y_2 = \dfrac{1}{2}t - 1$

Will the ants collide at any point in the future?

If the ants will collide, there must be a time $t$ when $x_1 = x_2$ and $y_1 = y_2$.

In this instance, that means $2t - 2 = t - 4$ and $4t - 3 = \frac{1}{2}t - 1$

Solving the first equation yields $t = -2$. Since time cannot go backwards, there is no real-world solution of the problem. The ants will not collide. We need not check whether the solution of the first equation would also be a solution of the second.

Parametric equations can also be used to draw figures. A circle, for instance, can be drawn with the equations

$x = \cos t$
$y = \sin t$

as $t$ increases from 0 to $2\pi$.

## 2.17 Finding powers of a binomial (the binomial theorem)

The figure below is known as **Pascal's triangle**. Only a portion of it is shown here. The left and right borders are all 1's. In the interior, each number is the sum of the two numbers above it to left and right.

```
Row 0:                    1
Row 1:                  1   1
Row 2:                1   2   1
Row 3:              1   3   3   1
Row 4:            1   4   6   4   1
Row 5:          1   5  10  10   5   1
Row 6:        1   6  15  20  15   6   1
```

Pascal's triangle can be used to find powers of a binomial without repeatedly multiplying. Notice that each row has a number, starting with row 0. The numbers in each row are the number coefficients of the terms of $(a + b)$ raised to the power of the row number. The variables in the terms start with $a^n$, followed by $a^{n-1}b$, $a^{n-2}b^2$, and so on, with decreasing exponents of $a$ and increasing exponents of $b$, ending with $ab^{n-1}$ and $b^n$. Putting the coefficients from row $n$ together with the variable exponents in the series produces the binomial expansion of $(a + b)^n$. For example: $(a + b)^3 = a^3 + 3a^2b + 3ab^2 + b^3$

Any binomial can be expanded in this fashion by setting its two terms equal to $a$ and $b$.

**Example:** Find $(2x - 3)^4$.

$a = 2x, b = -3$

$$(a+b)^4 = a^4 + 4a^3b + 6a^2b^2 + 4ab^3 + b^4$$
$$= (2x)^4 + 4(2x)^3(-3) + 6(2x)^2(-3)^2 + 4(2x)(-3)^3 + (-3)^4$$
$$= 16x^4 - 96x^3 + 216x^2 - 216x + 81$$

## 2.18 Composition of functions

**Composition of functions** is a way of combining functions such that the range of one function is the domain of another. For instance, the composition of functions $f$ and $g$ can be either $f \circ g$ (the composite of $f$ with $g$) or $g \circ f g + f$ (the composite of $g$ with $f$). Another way of writing these compositions is $f(g(x))$ and $g(f(x))$. The domain of the composition $f(g(x))$ includes all values $x$ such that $g(x)$ is in the domain of $f(x)$.

**Example:** What is the composition $f \circ g$ for the functions $f(x) = ax$ and $g(x) = bx^2$?

The correct answer can be found by substituting the function $g(x)$ into $f(x)$:

$$f(g(x)) = a \cdot g(x) = abx^2$$

On the other hand, the composition $g \circ f$ would yield a different answer.

$$g(f(x)) = b \cdot (f(x))^2 = b(ax)^2 = a^2bx^2$$

## 2.19 Inverses of functions

The **inverse** of a function $f(x)$ is typically labeled $f^{-1}(x)$ and satisfies the following two relations:

$$f(f^{-1}(x)) = x$$
$$f^{-1}(f(x)) = x$$

For a function $f(x)$ to have an inverse, it must be one-to-one. This fact is easily seen, since both $f(x)$ and $f^{-1}(x)$ must satisfy the vertical line test (that is, both must be functions). A function takes each value in a domain and relates it to only one value in the range. Logically, then, the inverse must do the same, only backwards: relate each value in the range to a single value in the domain.

## 2.20 Finding inverses of functions

Finding the inverse of a function can be a difficult or impossible task, but there are some simple approaches that can be followed in many cases. The simplest method for finding the inverse of a function is to interchange the variable and the function symbols and then solve to find the inverse. The approach is summarized in the outline below, given a one-to-one function $f(x)$.

1. Replace the symbol $f(x)$ with $x$
2. Replace all instances of $x$ in the function definition with $f^{-1}(x)$ (or $y$ or some other symbol)
3. Solve for $f^{-1}(x)$.
4. Check the result using $f(f^{-1}(x)) = x$ or $f^{-1}(f(x)) = x$.

**Example:** Determine if the function $f(x) = x^2$ has an inverse. If so, find the inverse.

First, determine if $f(x)$ is one-to-one. Note that $f(1) = f(-1) = 1$, so $f(x)$ is not one-to-one and therefore has no inverse function.

**Example:** Determine if the function $f(x) = x^3 + 1$ has an inverse. If so, find the inverse.

The function $f(x) = x^3 + 1$ has an inverse because it increases monotonically for $x > 0$ and decreases monotonically for $x < 0$. As a result, it is one-to-one, and the inverse exists. To calculate the inverse, let $y$ be $f^{-1}(x)$. Replace $f(x)$ with $x$ and replace $x$ with $y$.

$$f(x) = x^3 + 1 \rightarrow x = y^3 + 1$$

Solve for $y$.

$$x - 1 = y^3$$

$$y = \sqrt[3]{x-1}$$

$$f^{-1}(x) = \sqrt[3]{x-1}$$

Test the result.

$$f^{-1}(f(x)) = \sqrt[3]{(x^3+1)-1}$$
$$= \sqrt[3]{x^3+1-1} = \sqrt[3]{x^3} = x$$

The result is thus correct.

## 2.21 Operations with radicals

**Radicals** are inverse operators of exponents: $\sqrt[n]{a}$ is called the "$n$th root of $a$" and means the number that would have to be multiplied by itself $n$ times to produce $a$: $x = \sqrt[2]{a}$ means that $x \cdot x = a$, $x = \sqrt[3]{a}$ means $x \cdot x \cdot x = a$, and so on. The number under the radical sign, in this case $a$, is called the **radicand**. The number above the radicand to the left is called the **index** or **root**. When the root is omitted, it is always assumed to be 2. That is, $\sqrt{x} = \sqrt[2]{x}$.

Every positive number has two square roots, one positive and one negative. The square root of 16, for instance, is either 4 or $-4$, since $(+4)(+4) = 16$ and $(-4)(-4) = 16$. We can write the two results together as $\pm 4$. $+4$ is called the principal square root of 16. In many problems, the principal square root is the only answer that makes sense.

**Example:** find the length of one side of a square room having an area of 16 square feet. Here the only answer is $+4$ ft, since a length of $-4$ ft is meaningless.

### Addition and subtraction of radicals

1. We can only add or subtract radicals that have the same index and the same radicand.

    **Example:** $5\sqrt[3]{2} - 3\sqrt[3]{2} = 2\sqrt[3]{2}$

2. If the radicand is raised to a power equal to the index, the root operation cancels out the power operation.

Example: $\sqrt[5]{7^5} = 7$

3. If the radicand is raised to a power *different* from the index, convert the radical to its exponential form and apply laws of exponents.

Example: $\sqrt[3]{a^4} = (a^4)^{\frac{1}{3}} = a^{\frac{4}{3}}$

**Multiplication and division of radicals**

1. *Multiplication:* If the indexes or roots are the same, just multiply the radicands and keep the same index.

Example: $\sqrt{3} \times \sqrt{8} = \sqrt{3 \times 8} = \sqrt{24}$

If the indexes or roots are not the same but the radicands are the same, convert each number to its exponent form and apply laws of exponents.

Example: $\sqrt{a}\left(\sqrt[3]{a}\right) = a^{\frac{1}{2}} a^{\frac{1}{3}} = a^{\left(\frac{1}{2}+\frac{1}{3}\right)} = a^{\frac{5}{6}}$

2. *Division:* to divide by a radical denominator, we must eliminate the radical from the denominator. We call this operation "rationalizing" the denominator.

*case 1:* If the denominator has a single square root, we multiply both the numerator and denominator by the denominator.

Example: $\dfrac{3}{\sqrt{2}} = \dfrac{3}{\sqrt{2}}\left(\dfrac{\sqrt{2}}{\sqrt{2}}\right) = \dfrac{3\sqrt{2}}{2}$

*case 2:* If the denominator has two terms, we multiply the denominator and the numerator by the conjugate of the denominator. The conjugate is produced by changing the sign between the two terms from plus to minus or from minus to plus.

Example: 
$\dfrac{3}{5+\sqrt{2}} = \dfrac{3(5-\sqrt{2})}{(5+\sqrt{2})(5-\sqrt{2})}$   Multiply numerator and denominator by the conjugate of the denominator.

$= \dfrac{3(5-\sqrt{2})}{5^2 - (\sqrt{2})^2}$   Difference of two squares.

$= \dfrac{15 - 3\sqrt{2}}{23}$   Simplify using distributive property.

In this example, the conjugate of $5+\sqrt{2}$ is $5-\sqrt{2}$.

## 2.22 Piecewise functions

In a **piecewise function**, different functions are specified for different intervals of the domain.

Example: $f(x) = \begin{cases} -2x & \text{if } x < 2 \\ 0.5x+1 & \text{if } x \geq 2 \end{cases}$

Piecewise functions can be continuous or discontinuous. A piecewise function is continuous for a certain interval if it is defined for every point in the interval and if it can be drawn with a single stroke.

A piecewise function is discontinuous during a certain interval if it is undefined at any point or if there is a jump in value between subintervals. The function shown below is discontinuous, though it is defined for every point in the domain of real numbers, because the value jumps from −4 to 2 at x = 2.

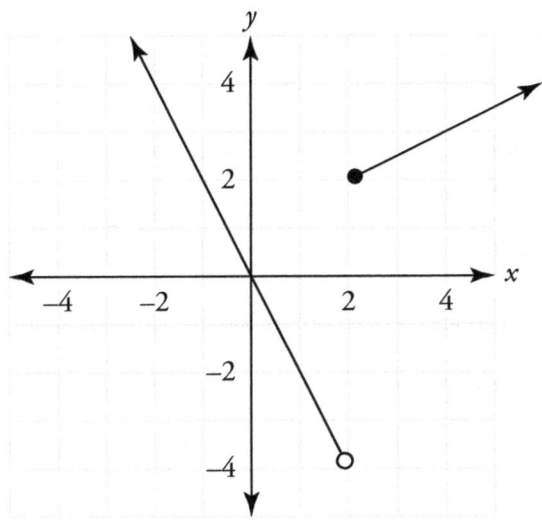

## 2.23 Inverse variation

In an **inverse variation**, two parameters vary in such a way that, as one gets larger, the other one gets smaller. If $x$ and $y$ vary inversely, there is a constant c such that $xy = c$.

Example: the greater the speed at which you drive, the shorter the time it takes to get to Grandma's house, 120 miles away. This is an inverse variation. If the speed in miles per hour is plotted as $x$ and the driving time in hours is plotted as $y$, then $xy = 120$, as in the graph below.

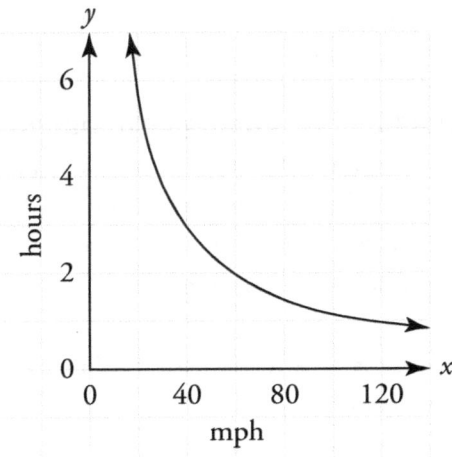

Example: If $30 were paid for 5 hours work, how much would be paid for 19 hours work? This is direct variation and $30 = 5c$, so the constant $c$ is 6 ($6/hour). So $y = 6(19) = $114$.

This could also be done as a proportion:

$$\frac{\$30}{5} = \frac{y}{19}$$
$$5y = \$570$$
$$y = \$114$$

## 2.24 Rational functions

A **rational function** $r(x)$ can be written as the ratio of two polynomial expressions $p(x)$ and $q(x)$, where $q(x)$ is nonzero.

$$r(x) = \frac{p(x)}{q(x)}, q(x) \neq 0$$

Examples of rational functions (and their associated expressions) are

$$r(x) = \frac{x^2 + 2x + 4}{x - 3} \text{ and } r(x) = \frac{x}{x^2 + 1}$$

Each of these examples is clearly the ratio of two polynomials. The following, however, is also a rational expression.

$$f(x) = \frac{1}{x + \frac{2}{x}}$$

This function can be shown to be a rational expression by converting it to standard form.

$$f(x) = \frac{1}{x + \frac{2}{x}}\left(\frac{x}{x}\right) = \frac{x}{x^2 + 2}$$

Since rational functions involve a denominator that is a polynomial expression (and not simply a constant), complicated division may be required to evaluate the function. Rational expressions are just like fractions and can be changed into other equivalent fractions through similar methods.

### Graphing rational functions using asymptotes

A function may have one or more asymptotes. An **asymptote** is a line for which the distance between it and a function or curve is arbitrarily small, especially as the function tends toward infinity in some direction. Asymptotes can be either vertical, horizontal or slant. Consider, for instance, the plot of the hyperbola defined as follows.

$$g(x) = \pm\sqrt{x^2 + 1}$$

Note, for instance, that as $x$ tends toward infinity, $g(x)$ gets arbitrarily close to $x$. The graph of $g(x)$ is shown below.

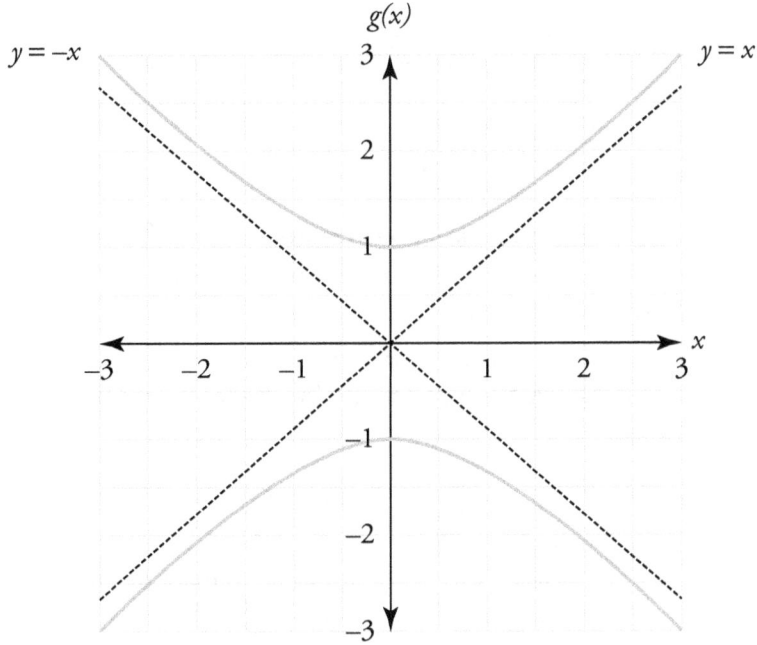

The (slant) asymptotes and their associated functions for this relation are displayed in the graph above as dashed lines.

Intervals of increase or decrease for a function are those regions over which the function is continuously increasing or decreasing, respectively. An interval of increase for a function $f(x)$ corresponds to any subset of the domain in which the slope of the function is always greater than zero; thus, if $x_2 > x_1$, then $f(x_2) > f(x_1)$ for any $x_1$ and $x_2$ in that interval of the function's domain. Likewise, an interval of decrease corresponds to any subset of the domain in which the slope of $f$ is less than zero.

## 2.25 Rational expressions

A **rational expression** is an expression having a variable in the denominator.

### Operations on rational expressions

The rules for operating on rational expressions are the same as those for numerical fractions.

*Addition and subtraction:* To add and subtract rational expressions, find a lowest common denominator (LCD), rewrite all the expressions using the LCD, then add or subtract and simplify if possible.

**Example:** $\dfrac{1}{x+4} - \dfrac{3}{x-1}$

$\text{LCD} = (x+4)(x-1)$

$\dfrac{1}{x+4} - \dfrac{3}{x-1} = \dfrac{1(x-1)}{(x+4)(x-1)} - \dfrac{3(x+4)}{(x+4)(x-1)}$  Rewrite expressions using LCD.

$\phantom{\dfrac{1}{x+4} - \dfrac{3}{x-1}} = \dfrac{x-1}{(x+4)(x-1)} - \dfrac{3x+12}{(x+4)(x-1)}$  Simplify.

Chapter 2: Algebra and Functions

$$= \frac{x-1-(3x+12)}{(x+4)(x-1)}$$ Subtract numerators over common denominator.

$$= \frac{-2x-13}{(x+4)(x-1)}$$ Simplify.

*Multiplication:* As with fractions, rational expressions can be multiplied simply by multiplying the numerators and multiplying the denominators.

**Example:** $\dfrac{2a}{a+b}\left(\dfrac{a-b}{c}\right) = \dfrac{2a(a-b)}{(a+b)c} = \dfrac{2a^2 - 2ab}{ac+bc}$

*Division:* The rule is the same as for fractions: invert the divisor (the second term) and multiply.

**Example:** $\dfrac{a^2-a}{b^3} \div \dfrac{a-1}{b} = \dfrac{a^2-a}{b^3} \times \dfrac{b}{a-1}$  Invert divisor and multiply.

$$= \frac{a(a-1)b}{b^3(a-1)}$$ Multiply numerators, multiply denominators.

$$= \frac{a}{b^2}$$ Cancel common factors from numerator and denominator.

## Simplifying rational expressions

To reduce a rational expression with more than one term in the denominator, the expression must be factored first. Factors that are the same will cancel. Addition or subtraction of rational expressions may first require finding a common denominator.

The first step to this end is to factor the denominators of both expressions to find the common factors. Then, proceed to rewrite the expressions with the common denominator by using the same methods as are used for numerical fractions.

**Example:** Rewrite the following fraction with a denominator of $(x+3)(x-5)(x+4)$:

$$\frac{x+2}{x^2+7x+12}$$

First, factor the denominator.

$$\frac{x+2}{x^2+7x+12} = \frac{x+2}{(x+3)(x+4)}$$

Multiply both the numerator and denominator by $(x-5)$:

$x$ 1 2

$$\frac{x+2}{x^2+7x+12} = \frac{x+2}{(x+3)(x+4)}\left(\frac{x-5}{x-5}\right) = \frac{(x+2)(x-5)}{(x+3)(x-5)(x+4)}$$

Although it is not necessary, the numerator and denominator can be multiplied out to represent the result as a rational expression in terms of polynomials.

$$\frac{x+2}{x^2+7x+12} = \frac{x^2-3x-10}{x^3+2x^2-23x-60}$$

The use of common denominators is helpful for addition and subtraction of rational expressions. Multiplication and division of rational expressions follows the standard rules of those operations.

Example: Evaluate the following expression.

$$\frac{5}{x^2-9} - \frac{2}{x^2+4x+3}$$

Let the expression above be labeled $f(x)$. First, find the common denominator, then subtract appropriately.

$$f(x) = \frac{5}{(x+3)(x-3)} - \frac{2}{(x+3)(x+1)}$$
$$= \frac{5(x+1)}{(x+3)(x-3)(x+1)} - \frac{2(x-3)}{(x+3)(x-3)(x+1)}$$
$$= \frac{5(x+1) - 2(x-3)}{(x+3)(x-3)(x+1)} = \frac{5x+5-2x+6}{(x+3)(x-3)(x+1)}$$
$$= \frac{3x+11}{(x+3)(x-3)(x+1)} = \frac{3x+11}{x^3+x^2-9x-9}$$

The above expression is the result, both in factored form and in standard form.

Example: Evaluate the following expression.

$$\frac{x^2-2x-24}{x^2+6x+8}\left(\frac{x^2+3x+2}{x^2-13x+42}\right)$$

Label the expression as $f(x)$. First, factor each polynomial, simplifying as appropriate, then multiply.

$$f(x) = \frac{(x-6)(x+4)}{(x+2)(x+4)} \frac{(x+1)(x+2)}{(x-6)(x-7)}$$
$$= \frac{x+1}{x-7}$$

## Solving equations involving rational expressions

To solve an equation containing rational expressions, set the expression equal to zero (which leads to the elimination of the denominator) and solve, as with simple polynomials.

$$r(x) = \frac{p(x)}{q(x)}$$

Note, however, that solutions to $p(x) = 0$ may lead to undefined values for $r(x)$, (that is, values for which $q(x) = 0$), and must be checked prior to acceptance.

This difficulty can be alleviated to some extent by factoring $p(x)$ and $q(x)$ and eliminating common factors.

**Example:** Find the solutions for $\dfrac{12}{2x^2 - 4x} + \dfrac{13}{5} = \dfrac{9}{x-2}$

Factor and rearrange the equation as follows, then solve for $x$.

$$\dfrac{12}{2x(x-2)} - \dfrac{9}{x-2} = -\dfrac{13}{5}$$

$$\dfrac{12}{2x(x-2)} - \dfrac{9(2x)}{2x(x-2)} = \dfrac{-18x+12}{2x(x-2)} = -\dfrac{13}{5}$$

$$-18x + 12 = -\dfrac{13}{5}(2x)(x-2) = -\dfrac{26}{5}x^2 + \dfrac{52}{5}x$$

$$\dfrac{26}{5}x^2 - \dfrac{52}{5}x - 18x + 12 = 0$$

$$0 = \dfrac{26}{5}x^2 - \dfrac{142}{5}x + 12 = 26x^2 - 142x + 60 = 13x^2 - 71x + 30$$

The solutions for $x$ can be found by factoring the above expression.

$$13x^2 - 71x + 30 = (x-5)(13x-6) = 0$$

Thus, $x = 5$ or $x = \dfrac{6}{13}$. These solutions can be confirmed by substitution into the original equation.

## 2.26 Polynomial functions

A **polynomial** is a sum of terms, where each term is a constant multiplied by a variable raised to a positive integer power. The general form of a polynomial $P(x)$ is:

$$a_n x^n + a_{n-1} x^{n-1} + \ldots + a_2 x^2 + a_1 x + a_0$$

Polynomials written in standard form have the terms written in order of decreasing exponent value, as shown above. The degree of a polynomial function in one variable is the value of the largest exponent to which the variable is raised. The above expression is a polynomial of degree $n$ (assuming that $a_n \neq 0$). Any function that represents a line, for instance, is a polynomial function of degree 1. Quadratic functions are polynomials of degree 2.

There are many methods for solving problems that involve polynomial equations.

For instance, in cases in which a polynomial is highly complicated or involves constants that do not permit methods such as factoring, a numerical approach may be appropriate. Newton's method is one possible approach to solving a polynomial equation numerically. At other times, solving a polynomial equation may require a graphical approach whereby the behavior of the function is examined on a visual plot. When using Newton's method, graphing the function can be helpful for estimating the locations of the roots (if any).

Polynomial equations with real coefficients cannot always be solved using only real numbers.

**Example:** Consider the quadratic function given below:

$$f(x) = x^2 + 1$$

There are no real roots for this equation, since

$x^2 + 1 = 0 \to x^2 = -1$.

The Fundamental Theorem of Algebra (see below), however, indicates that there must be two (possibly non-distinct) solutions to this equation. Note that if the complex numbers are permitted as solutions to this equation, then

$x = \pm i$.

Thus, generally, solutions to any polynomial equation with real coefficients exist in the set of complex numbers.

If a phenomenon or situation can be modeled with a polynomial equation, the following theorems can be helpful in solving the equation. These theorems include the Fundamental Theorem of Algebra, the Factor Theorem, the Complex Conjugate Root Theorem, and the Rational Root Theorem.

## 2.27 The fundamental theorem of algebra

The **fundamental theorem of algebra** states that a polynomial expression of degree $n$ must have $n$ roots (which may be real or complex and which may not be distinct). It follows from the theorem that if the degree of a polynomial is odd, then it must have at least one real root.

Polynomial functions are in the form of $P(x)$ given below, where $n$ is the degree of the polynomial and the constant $an$ is nonzero.

$$P(x) = a_n x^n + a_{n-1} x^{n-1} + \ldots + a_2 x^2 + a_1 x + a_0 \text{ asdfasdf}$$

If $P(c) = 0$ for some number $c$, then $c$ is said to be a *zero* (or *root*) of the function.

A zero is also called a *solution* to the equation $P(x) = 0$.

The existence of $n$ solutions can be seen by looking at a factorization of $P(x)$.

**Example:** Find the roots of $P(x) = x^2 - x - 6$.

This second-degree polynomial can be factored into

$P(x) = (x + 2)(x - 3)$.

Note that $P(x)$ has two roots in this case: $x = -2$ and $x = 3$. This corresponds to the degree of the polynomial, $n = 2$. In some cases, however, there may be nondistinct roots. Consider $P(x) = x^2$.

$P(x) = x(x)$

Note that the polynomial is factored in the same way as the previous example, but, in this case, the roots are identical: $x = 0$. Thus, although there are two roots for this second-degree polynomial, the roots are not distinct.

Likewise, roots of a polynomial may be complex.

**Example:** Find the roots of $P(x) = x^2 + 1$.

The range of this function is $P(x) \geq 1$, so there are no real roots, since the function never crosses the $x$-axis. Nevertheless, if complex values of $x$ are permitted, there are cases where $P(x)$ is zero. Factor $P(x)$ as before, but this time use complex numbers.

$$P(x) = (x + i)(x - i)$$

The solutions are $x = i$ and $x = -i$. Thus, this second-degree polynomial still has two roots.

## 2.28 The factor theorem

The **factor theorem** establishes the relationship between the factors and the zeros or roots of a polynomial and is useful for finding the factors of higher-degree polynomials. The theorem states that a polynomial $P(x)$ $(x - c)$ if and only if $P(c) = 0$.

For a general $n$th-degree polynomial, the function $P(x)$ can therefore be factored as follows:

$$P(x) = (x - c_n)(x - c_{n-1})...(x - c_2)(x - c_1)$$

As with the second-degree polynomial examples examined above, a general $n$th-degree polynomial can have roots $c_i$ that are distinct or nondistinct, and real or complex. Since this is the case, if all of the roots of a polynomial are known, then a function $P(x)$ is determined based on the factoring approach shown above.

In addition, if a single root $c$ is known, then the polynomial can be simplified (that is, it can be reduced by one degree) using division.

$$Q(x) = \frac{P(x)}{x - c}$$

Here, if $P(x)$ has degree $n$, then $Q(x)$ has degree $n - 1$. If some number of roots are known, the task of finding the remainder of the roots can be simplified by performing the division represented above. As each successive root is found, the degree of the polynomial can be reduced to further simplify finding the remainder of the roots.

## 2.29 The rational root theorem

The **rational root theorem**, also known as the rational zero theorem, allows determination of all possible rational roots (or zeroes) of a polynomial equation with integer coefficients. (A root is a value of $x$ such that $P(x) = 0$.) Every rational root of $P(x)$ can be written as $x = \frac{p}{q}$, where $p$ is an integer factor of the constant term $a_0$ and $q$ is an integer factor of the leading coefficient $a_n$.

**Example:** find the possible rational roots of $3x^3 - 7x^2 + 3x - 2$.

$3x^3 - 7x^2 + 3x - 2$

$p$ must be an integer factor of $-2$: 1, $-1$, 2, or $-2$.

$q$ must be an integer factor of 3: 1 or 3.

$$\frac{p}{q} = \frac{1, -1, 2, \text{ or } -2}{1 \text{ or } 3} = 1, -1, 2, -2, \frac{1}{3}, -\frac{1}{3}, \frac{2}{3}, \text{ or } -\frac{2}{3}$$

The rational root theorem guarantees that any rational roots of $3x^3 - 7x^2 + 3x - 2$ will be in the list just given, but it does NOT guarantee that every item in the list will be a root. Test each possibility. The only result that works is $x = 2$.

## 2.30 The complex conjugate root theorem

The **complex conjugate root theorem** states that for a polynomial $P(x)$ with real coefficients, if $P(x)$ has a complex root $a + bi$, then it must also have a complex root $a - bi$. In the case of quadratic equations, this becomes obvious when the quadratic formula is used, since the imaginary part of any root is always calculated as "plus or minus." However, it also holds for higher-power polynomials.

## 2.31 Quadratic inequalities

To solve a quadratic inequality, first gather all terms on the left side of the inequality sign, leaving zero on the right.

**Example:** $x^2 < 5x + 6 \rightarrow x^2 - 5x - 6 < 0$

Then replace the sign of the inequality with an equals sign and solve the resulting quadratic equation by factoring or by using the quadratic formula:

$$x^2 - 5x - 6 = 0 \rightarrow (x+1)(x-6) = 0 \rightarrow x = -1, 6$$

The last step is to restore the inequality signs. This can be done by graphing the quadratic equation as a function. In this example, since the first term of the equation is positive, the resulting graph will be a parabola opening upward, with zeros at $-1$ and $6$.

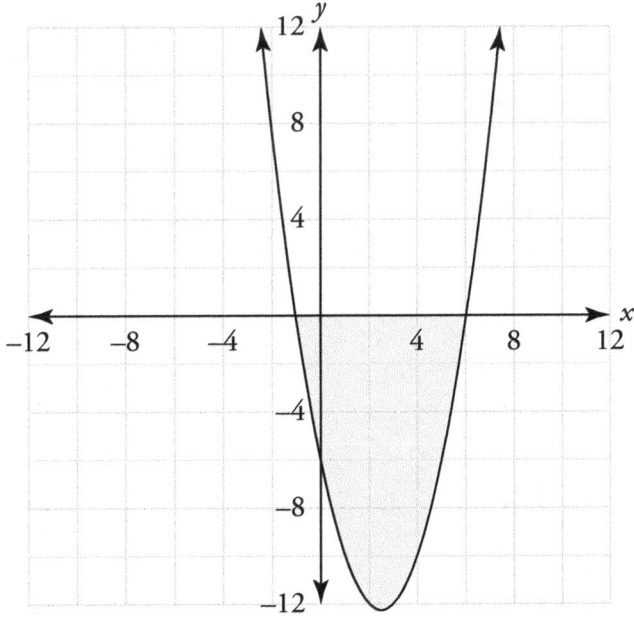

To restore the inequality signs, note that in this example, the solutions are supposed to be less than zero. Those solutions will be found in the part of the parabola that goes below the *x*-axis, namely the section between −1 and 6. Therefore, the solutions are $-1 < x < 6$.

## 2.32 Modeling functions

Functions can be represented in a variety of ways, including as a symbolic expression (for instance, $f(x) = 3x^2 - \sin x$), a graph, a table of values, and a common-language expression (for example, "the speed of the car increases linearly from 0 to 100 miles per hour in 12 seconds"). The ability to convert among various representations of a function depends on how much information is provided. For instance, although a graph of a function can provide some clues as to its symbolic representation, it is often difficult or impossible to obtain an exact symbolic form based only on a graph. The same difficulty applies to tables.

Converting from a symbolic form to a graph or table, however, is relatively simple, especially if a computer is available. The symbolic expression need simply be evaluated for a representative set of points that can be used to produce a sufficiently detailed graph or table. For example, the equation $y = 9x$ describes the relationship between *y*, the total number of dollars earned, and *x*, the number of $9 sunglasses sold. In a relationship of this type, one of the quantities (e.g., total amount earned) is dependent on the other (e.g., number of sunglasses sold). These variables are known as the dependent and independent variables, respectively. A table using this data would appear as:

| number of sunglasses sold | 1 | 5 | 10 | 15 |
|---|---|---|---|---|
| total dollars earned | 9 | 45 | 90 | 135 |

Each (*x*, *y*) relationship between a pair of values is called a coordinate pair and can be plotted on a graph. The coordinate pairs (1, 9), (5, 45), (10, 90), and (15, 135) are plotted on the graph below.

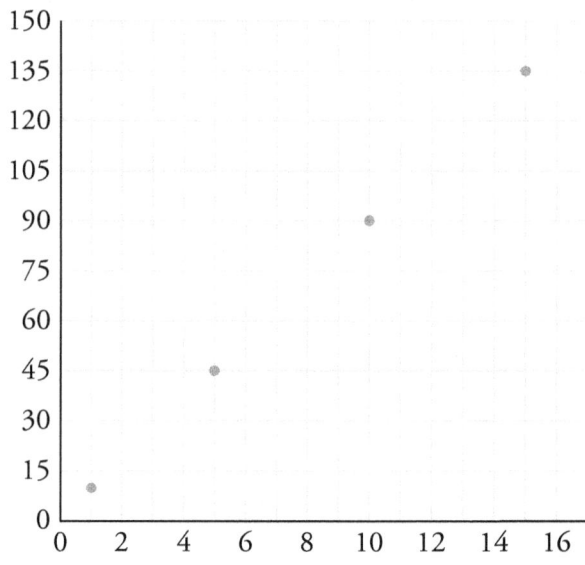

The graph shows a linear relationship. A **linear relationship** is one in which the change in two quantities is in a constant proportion. Doubling the change in $x$ also doubles the change in $y$. On a graph, a straight line depicts a linear relationship.

The function or relationship between two quantities may be analyzed to determine how one quantity depends on the other. For example, the function $y = 2x + 1$ below shows a relationship between $y$ and $x$.

The relationship between two or more variables can be analyzed using a table, graph, written description or symbolic rule. The function $y = 2x + 1$ is written as a symbolic rule. The same relationship is also shown in the table below:

| $x$ | 0 | 2 | 3 | 6 | 9 |
|---|---|---|---|---|---|
| $y$ | 1 | 5 | 7 | 13 | 19 |

A relationship could be written in words by saying the value of $y$ is equal to two times the value of $x$, plus one. This relationship could be shown on a graph by plotting given points such as the ones shown in the table above.

Another way to describe a function is as a process in which one or more numbers are input into an imaginary machine that produces another number as the output. If 3 is input as $x$ into a machine with a process of $2x + 1$, the output, $y$, will equal 7.

In real situations, relationships can be described mathematically. The function $y = x + 1$ can be used to describe the idea that people age one year on their birthday. To describe the relationship in which a person's monthly medical costs are 6 times a person's age, we could write $y = 6x$. The monthly cost of medical care could be predicted using this function. A 20-year-old person would spend $120 per month ($120 = 20 \times 6$). An 80-year-old person would spend $480 per month ($480 = 80 \times 6$). Therefore, one could analyze the relationship to say: as you get older, monthly medical costs increase $6.00 each year.

**Modeling a linear function from a table**

**Example:** What is the equation that expresses the relationship between $x$ and $y$ in the table below?

| $x$ | 0 | 1 | 2 | 3 | 4 | 5 |
|---|---|---|---|---|---|---|
| $y$ | 3 | 5 | 7 | 9 | 11 | 13 |

Each $x$-input differs from the next by a constant, 1. Since each $y$-output also differs from the next by a constant, 2, these data can be modeled by a linear function.

We will write the equation as $y = mx + b$. We can determine $m$ (the slope) as the change in $y$ between any two points divided by the change in $x$ between any two points. In this table, the change in $y$ between two points is 2 and the change in $x$ between two points is 1, so the slope $m$ is $\frac{2}{1} = 2$. Find $b$ by setting $x$ to zero. In this table, when $x = 0$, $y = 3$, so $b = 3$, and the equation is $y = 2x + 3$.

## Modeling a quadratic function from a table

A **quadratic function** can also be detected from data arranged in a table, particularly if the $x$-inputs differ by a constant, preferably 1. In the cases of data generated by a quadratic function, the difference between one output and the next will not be a constant, but the difference between one difference and the next, the so-called "second difference," will be constant.

Example:

| $x$ | $y$ | 1st Difference | 2nd Difference |
|---|---|---|---|
| 0 | −1 |  |  |
|  |  | 1 |  |
| 1 | 0 |  | 4 |
|  |  | 5 |  |
| 2 | 5 |  | 4 |
|  |  | 9 |  |
| 3 | 14 |  | 4 |
|  |  | 13 |  |
| 4 | 27 |  |  |

Since the second differences are constant, the data in this table could be modeled by a quadratic function.

## Modeling an exponential function from a table

An exponential function can also be detected from a table in which the $x$-inputs differ by a constant value (preferably 1). In the case of an **exponential function**, the ratio of each item to the next will be a constant.

Example:

| $x$ | $y$ | Ratio |
|---|---|---|
| 0 | 1 |  |
|  |  | ×1.5 |
| 1 | 1.5 |  |
|  |  | ×1.5 |
| 2 | 2.25 |  |
|  |  | ×1.5 |
| 3 | 3.375 |  |
|  |  | ×1.5 |
| 4 | 5.0625 |  |

Since the ratio of each item to the previous one is constant, the data above can be modeled by an exponential function.

# Chapter 3: Coordinate Geometry

## 3.1 Graphing linear equations

The graph of a linear equation represents a straight line. It takes two points to define a unique straight line.

1. Choose only 3 values of $x$.
2. Substitute each chosen value of $x$ in the equation to find the corresponding $y$-value.
3. Plot the 3 points and join them with a straight line.

**Intercepts**

The **intercepts** of a function are the points at which the function crosses the $x$- or $y$-axis. Since the $x$-value of any point on the $y$-axis is 0, the $y$-intercept of any function can be found by setting $x$ equal to 0 and using the function to find the corresponding $y$-value.

**Example:** find the $y$-intercept of $f(n) = x^2 + x + 4$

Let $x = 0$. $f(0) = 0^2 + 0 + 4 = 4$.

The $y$-intercept is at $(0, 4)$.

Similarly, setting $f(x)$ equal to 0 and solving for $x$ makes it possible to find an $x$-intercept. Whereas a function normally has one $y$-intercept, a function can have 0, 1, or multiple $x$-intercepts.

**Example:** find any $x$-intercepts for the function $f(x) = x^2 - 25$

Let $f(x) = 0$. $0 = x^2 - 25$

$x^2 = 25$

$x = \pm 5$

There are $x$-intercepts at $(5, 0)$ and $(-5, 0)$.

**Note:** It is typically helpful to choose the $x$-intercept and the $y$-intercept as the two key points (when possible) when graphing a line.

**Example:** sketch the graph of the line represented by

$2x + 3y = 6$

Let $x = 0 \Rightarrow 2(0) + 3y = 6$

$\Rightarrow 3y = 6$

$\Rightarrow y = 2$

$\Rightarrow (0, 2)$ is the $y$-intercept.

Let $y = 0 \Rightarrow 2x + 3(0) = 6$

$\Rightarrow 2x = 6$

$\Rightarrow x = 3$

$\Rightarrow (3, 0)$ is the $x$-intercept.

Let $x = 1 \Rightarrow 2(1) + 3y = 6$

$\Rightarrow 2 + 3y = 6$ (subtract 2 from both sides)

$\Rightarrow 3y = 4$ (Divide both sides by 3)

$\Rightarrow y = \frac{4}{3}$

$\Rightarrow \left(1, \frac{4}{3}\right)$ is the third point.

Plotting the 3 points or the coordinate system, we get the following graph:

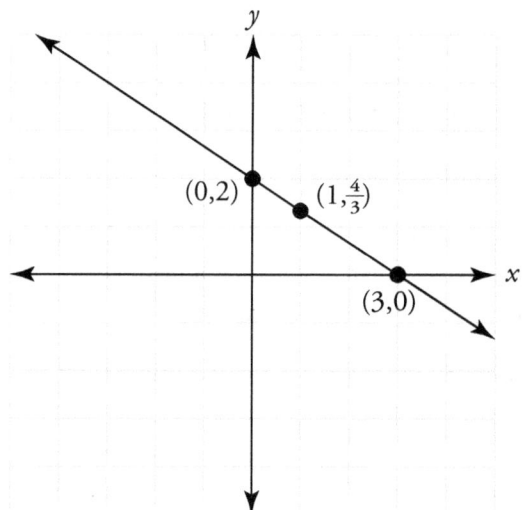

**Note:** Two points are sufficient to graph the line; the third point is for checking purposes.

### Solving a system of equations by graphing

A system of equations in two variables can be solved by solving each equation for $y$ and graphing each equation on a common set of axes. The intersection point is the solution for $x$ and $y$.

Example: solve by graphing.

$$\begin{cases} 2x - y = 1 & [1] \\ x + y = 2 & [2] \end{cases}$$

Solving [1] for $y$ gives $y = 2x - 1$. Solving [2] for $y$ gives $y = -x + 2$. Plot the two lines.

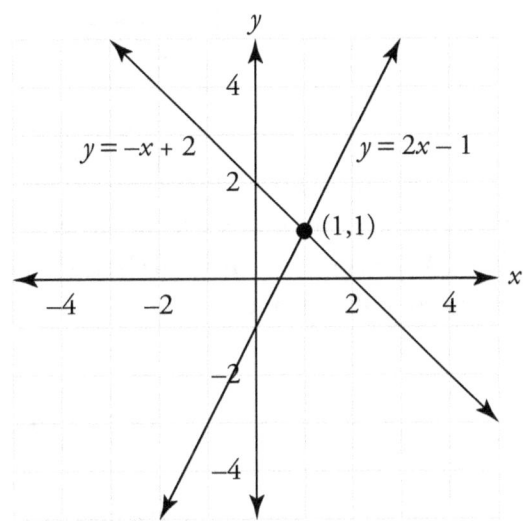

The intersection point appears to be at (1, 1), indicating a solution of $x = 1$, $y = 1$.

Because solving by graphing depends on the accuracy of visual inspection, solving by elimination or substitution is more reliable.

## 3.2 Graphing quadratic equations

The graph of a quadratic function in the form $y = ax^2 + bx + c$ is a **parabola**. If the square term is positive, the parabola opens upward. If the square term is negative, the parabola opens downward.

Example: The graph of $y = 2x^2 - 2x - 2$ opens upward. The graph of $y = -x^2 + 3x + 3$ opens downward.

If the function is written so that $x$ is a function of $y$, then the parabola will open to the right if the square term is positive and will open to the left if that term is negative.

Example: The graph of $x = y^2 + 1$ opens to the right. The graph of $x = -y^2 - y + 2$ opens to the left.

### Graphing a parabola

Make a table of several negative and positive values of $x$ and compute the $y$-value for each $x$-value. Plot the resulting coordinate pairs and draw a smooth curve through the points.

**Example:** Graph $-\frac{1}{2}x^2 + 2x + 2$.

| $x$ | $y$ |
|---|---|
| −3 | −8.5 |
| −2 | −4 |
| 0 | 2 |
| 2 | 4 |
| 4 | 2 |
| 6 | −4 |

Notice the same $y$-values for $x = 0$ and $x = 4$ and the same $y$-values for $x = -2$ and $x = 6$. A parabola is symmetrical about its vertical axis, and these duplicate values represent that symmetry.

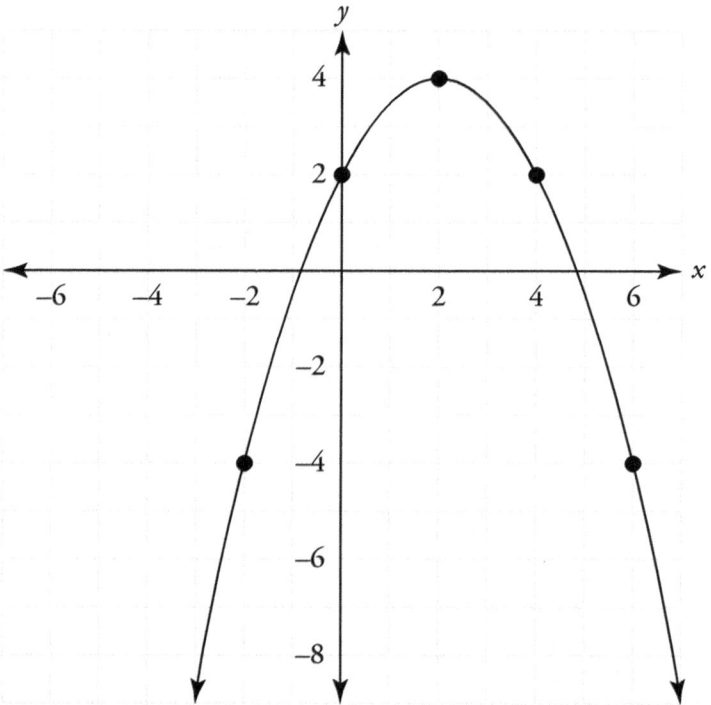

### Vertex of a parabola

Every parabola has either a maximum or a minimum point called a **vertex**. If the parabola opens up, the vertex is a minimum. If the parabola opens down, the vertex is a maximum.

To find the $x$-value of the vertex of a parabolic function, use the equation $x = \frac{-b}{2a}$. (This equation is also the equation of the parabola's **axis of symmetry**.) Then substitute the $x$-value into the original function to find the $y$-value of the vertex.

**Example:** Find the axis of symmetry and the vertex of the parabola corresponding to the function $y = x^2 + 8x + 15$

$$x = \frac{-b}{2a} = \frac{-8}{2(1)} = -4$$

$$\begin{aligned} y &= x^2 + 8x + 15 \\ &= (-4)^2 + 8(-4) + 15 \\ &= 16 - 32 + 15 \\ &= -1 \end{aligned}$$

$(x, y) = (-4, -1)$

The axis of symmetry is at $x = -4$. The vertex is at $(-4, -1)$.

## Circles

The equation $x^2 + y^2 = r^2$ draws a circle of radius $r$ around the origin. In the figure below it is clear from the Pythagorean Theorem that $x^2 + y^2 = r^2$. A **circle**, then, consists of all the points for which $r$ is a constant value.

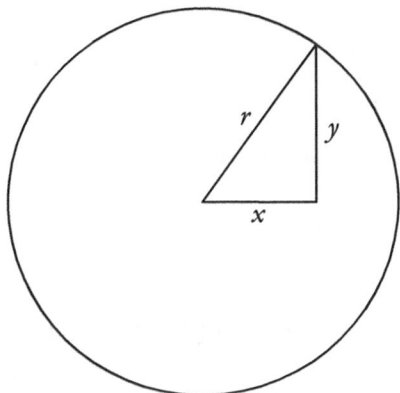

**Example:** Draw a circle of radius 5 around the origin.

$$x^2 + y^2 = r^2$$
$$x^2 + y^2 = 5^2$$
$$x^2 + y^2 = 25$$

## Ellipses

In a circle, all points are a fixed distance from a single point, the center. An ellipse is not defined by a single center point but by two central points called foci. For every point on an **ellipse**, the sum of the distances from the two foci is the same. This definition can be made clear by sticking two pushpins in a piece of cardboard and looping a length of string around them. If you keep a pencil point to keep the loop taut as you move around the pins, the pencil will draw an ellipse.

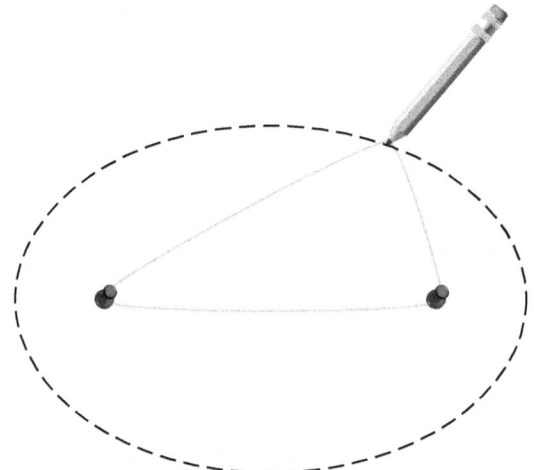

Instead of a uniform diameter, an ellipse has a **major axis** and a **minor axis**. The major axis extends from one end of the ellipse through the foci to the other end. The minor axis is perpendicular to the major axis. The equation for an ellipse centered at the origin is

$$\frac{x^2}{a^2} + \frac{y^2}{b^2} = 1$$

If $a^2 > b^2$, the major axis is horizontal. If $b^2 > a^2$, the major axis is vertical. The lengths of the two axes are $2a$ and $2b$, respectively.

**Example:** Describe the ellipse corresponding to

$$\frac{x^2}{36} + \frac{y^2}{64} = 1$$

$a^2 = 36$, $b^2 = 64$. Since $b^2$ is larger, the major axis is vertical and $2\sqrt{64} = 16$ units long.

The minor axis is horizontal and is $2\sqrt{36} = 12$ units long.

**Example:** Write an equation that would produce an ellipse centered at the origin with a horizontal axis 14 units long and a vertical axis 6 units long.

Since 14 units represents $2a$, $a = 7$. Since 6 units $= 2b$, $b = 3$. The required equation is

$$\frac{x^2}{a^2} + \frac{y^2}{b^2} = 1$$

$$\frac{x^2}{49} + \frac{y^2}{9} = 1$$

## Hyperbolas

A **hyperbola** is a curve—actually a pair of curves—that is generated, like an ellipse, by picking two points to act as foci. For every point on a hyperbola, the difference in the distances from the two foci is the same. An equation in the form $\frac{x^2}{a^2} - \frac{y^2}{b^2} = 1$ produces a hyperbola centered around the origin, opening to the left and right, with the foci lying on a horizontal line. An equation in the form $\frac{y^2}{a^2} - \frac{x^2}{b^2} = 1$ produces a hyperbola opening up and down, with the foci lying on a vertical line.

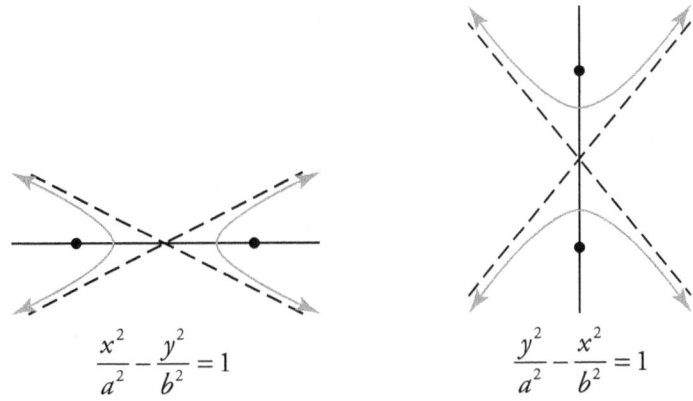

$$\frac{x^2}{a^2} - \frac{y^2}{b^2} = 1 \qquad \frac{y^2}{a^2} - \frac{x^2}{b^2} = 1$$

Hyperbolas are bounded by **asymptotes**, lines that the curves continually approach without reaching.

A hyperbola in the form $\frac{x^2}{a^2} - \frac{y^2}{b^2} = 1$ intercepts the x-axis at $(a, 0)$ and $(-a, 0)$. Its asymptotes are $y = \frac{b}{a}x$ and $y = -\frac{b}{a}x$.

A hyperbola in the form $\frac{y^2}{a^2} - \frac{x^2}{b^2} = 1$ intercepts the x-axis at $(0, a)$ and $(0, -a)$. Its asymptotes are $y = \frac{a}{b}x$ and $y = -\frac{a}{b}x$.

To graph the hyperbola, plot the two x- or y-intercepts. Sketch the asymptotes as dotted lines. Then sketch the two curves so that they pass through the intercepts and approach the asymptotes.

**Example:** Sketch the hyperbola defined by $\frac{y^2}{6.25} - \frac{x^2}{12.25} = 1$

$a = 2.5, b = 3.5$.

Intercepts at $(0, 2.5)$ and $(0, -2.5)$.

Asymptotes: $y = \pm\frac{2.5}{3.5}x = \pm\frac{5}{7}x$

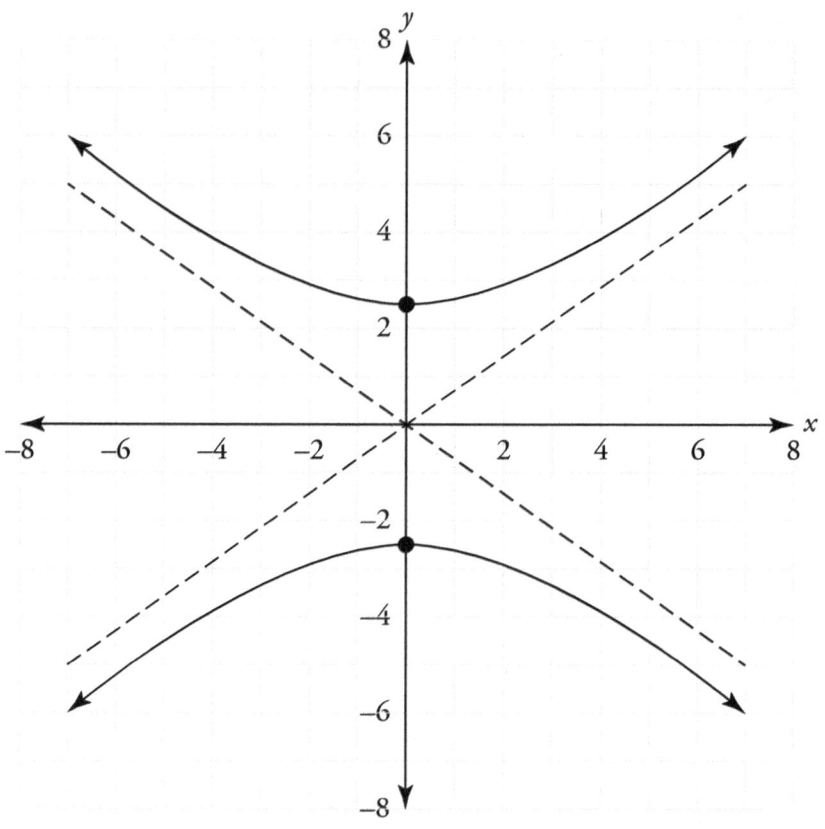

## 3.3 Transformational geometry: translations, rotations, reflections, and scaling

**Transformational geometry** is the study of the manipulation of objects through movement, rotation, and scaling. The transformed version of an object is called its **image**. If the original object is labeled with letters, such as ABCD, the image can be labeled with the same letters followed by a prime symbol: A′B′C′D′.

Transformations can be characterized in different ways.

### Types of transformations

An **isometry** is a linear transformation that maintains the dimensions of a geometric figure. Translations, rotations, and reflections are all isometries.

## 3.4 Translations

A **translation** is a transformation that "slides" an object a fixed distance in a given direction. The original object and its translation have the same shape and size, and they face in the same direction.

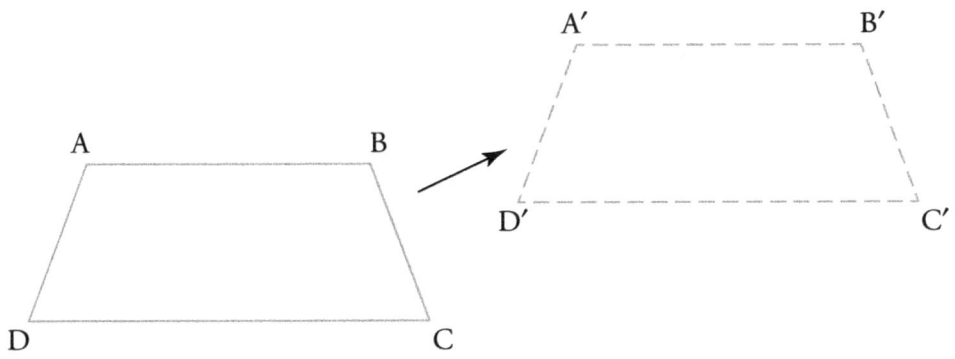

In the coordinate plane, a point or figure can be translated
- $h$ units to the right by adding $h$ to the x-coordinates of all points
- $h$ units to the left by subtracting $h$ from the x-coordinates of all points
- $k$ units up by adding $k$ to the y-coordinates of all points
- $k$ units down by subtracting $k$ from the y-coordinates of all points

Example: A triangle with coordinates $(-1, 2)$, $(3, 5)$ and $(1, -3)$ is translated 5 units left and 1 unit up. What are the coordinates of the translated triangle?

To translate 5 units left, 5 must be subtracted from each x-coordinate. To translate 1 unit up, 1 must be added to each y-coordinate. The coordinates of the new triangle are $(-6, 3)$, $(-2, 6)$ and $(-4, -2)$.

## Translating a parabola

The graph of a parabola produced by an equation in the form $y = ax^2 + bx + c$ can be translated
- $h$ units to the right by substituting $(x - h)$ for $x$
- $h$ units to the left by substituting $(x + h)$ for $x$
- $k$ units up by substituting $(y - k)$ for $y$
- $k$ units down by substituting $(y + k)$ for $y$

Example: Graph the function $y = x^2$. Then move it 3 units right, left, up and down.

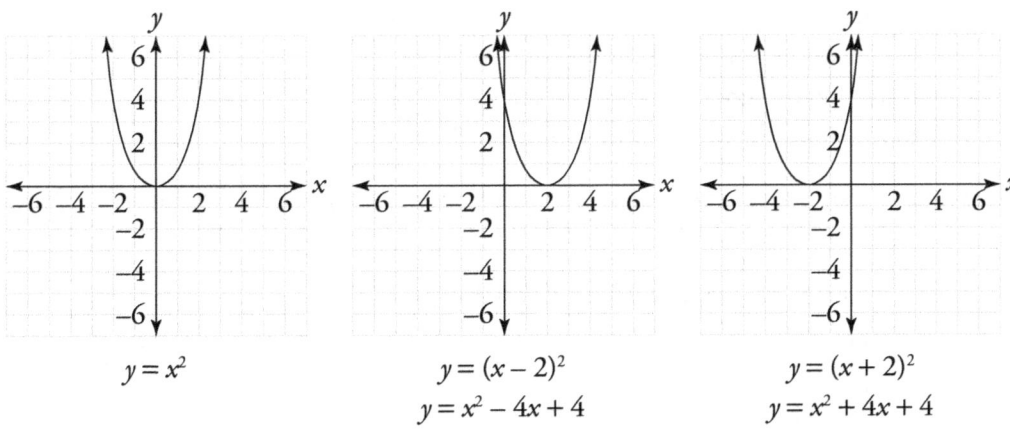

Chapter 3: Coordinate Geometry    81

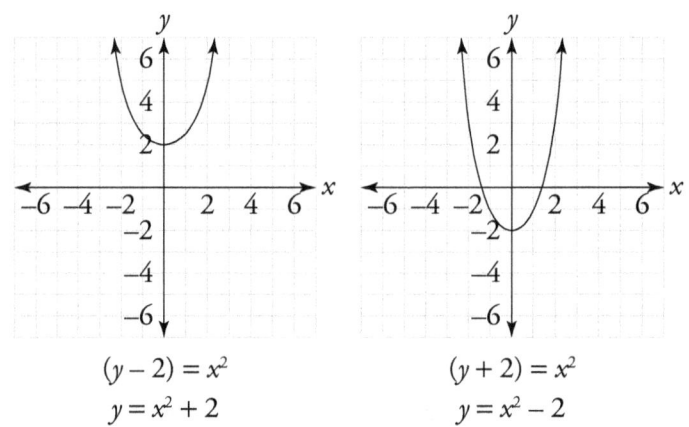

$(y-2) = x^2$　　　　$(y+2) = x^2$
$y = x^2 + 2$　　　　$y = x^2 - 2$

Summary: If $f(x)$ is a quadratic function producing a parabola, then $f(x-h) + k$ will produce a similar parabola shifted $h$ units to the right and $k$ units up.

### Translating a circle

The center of a circle can be translated
- $h$ units to the right by substituting $(x-h)$ for $x$
- $h$ units to the left by substituting $(x+h)$ for $x$
- $k$ units up by substituting $(y-k)$ for $y$
- $k$ units down by substituting $(y+k)$ for $y$

**Example:** Draw a circle of radius 4 with center at $(3, -2)$.

The center of the circle needs to be moved 3 units to the right and 2 units down.

$h = 3, k = -2, r = 4$
$(x-h)^2 + (y-k)^2 = r^2$
$(x-3)^2 + (y+2)^2 = 16$

### Translating an ellipse

An ellipse produced by an equation in the form $\dfrac{x^2}{a^2} + \dfrac{y^2}{b^2} = 1$ can be translated
- $h$ units to the right by substituting $(x-h)$ for $x$
- $h$ units to the left by substituting $(x+h)$ for $x$
- $k$ units up by substituting $(y-k)$ for $y$
- $k$ units down by substituting $(y+k)$ for $y$

**Example:** Write an equation of an ellipse with center at $(3, -2)$, horizontal axis of 9 and vertical axis of 12.

$2a = 9, a = 4.5, 2b = 12, b = 6, h = 3, k = -2$
$\dfrac{(x-h)^2}{a^2} + \dfrac{(y-k)^2}{b^2} = 1$
$\dfrac{(x-3)^2}{20.25} + \dfrac{(y+2)^2}{36} = 1$

**Translating a hyperbola**

A hyperbola in the form $\frac{x^2}{a^2} - \frac{y^2}{b^2} = 1$ can be translated
- $h$ units to the right by substituting $(x - h)$ for $x$
- $h$ units to the left by substituting $(x + h)$ for $x$
- $k$ units up by substituting $(y - k)$ for $y$
- $k$ units down by substituting $(y + k)$ for $y$

**Example:** Write the equation of a hyperbola with vertices at (2, 3) and (6, 3) and asymptotes at $y = \pm\frac{5}{2}(x - 4) + 3$. The distance between the vertices is $2a = 4$, so $a = 2$. From the equations of the asymptotes, $\frac{b}{a} = \frac{5}{2}$, so $b = 5$. The center is midway between the vertices at (4, 3), so $h = 4$, $k = 3$. The equation is

$$\frac{(x-h)}{a^2} - \frac{(y-k)}{b^2} = 1$$
$$\frac{(x-4)}{4} - \frac{(y-3)}{25} = 1$$

## 3.5 Rotations

A **rotation** is a transformation that turns a figure about a fixed point, which is called the center of rotation. An object and its rotation are the same shape and size, but the figures may be oriented in different directions. Rotations can occur in either a clockwise or a counterclockwise direction.

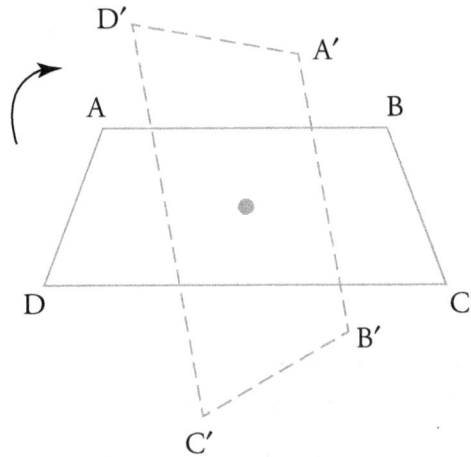

In the coordinate plane, a point or figure can be rotated about the origin
- 90° counterclockwise (or 270° clockwise) by changing $(x, y)$ to $(-y, x)$

  [**Example:** the point (−3, 2), rotated 90° counterclockwise around the origin, produces the point (−2, −3).]

- 180° clockwise or counterclockwise by changing $(x, y)$ to $(-x, -y)$

  [**Example:** the point (6, −7), rotated 180° around the origin, produces the point (−6, 7).]

- 90° clockwise (or 270° counterclockwise) by changing $(x, y)$ to $(y, -x)$

    [Example: the point $(-1, 5)$, rotated 90° clockwise around the origin, produces the point $(5, 1)$.]

Example: A triangle with coordinates $(-3, 2)$, $(4, -1)$, $(1, -2)$ is rotated 90° clockwise. What are the coordinates of the rotated triangle?

To produce a 90° clockwise rotation, each *x*-coordinate must be replaced with its opposite and the two coordinates must be switched.

The coordinates of the new triangle are $(2, 3)$, $(-1, -4)$, $(-2, -1)$.

## 3.6   Reflections

The **reflection** of a figure across a line, called the line of reflection, produces a figure similar but reversed, at an equal distance on the opposite side of the line, as if it were a mirror image and the line of reflection was the mirror.

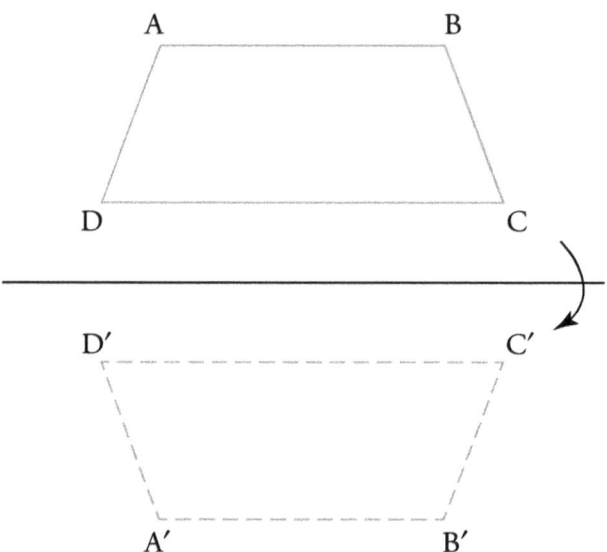

In the coordinate plane, a point or figure can be reflected
- across the *y*-axis by changing $(x, y)$ to $(-x, y)$

    [Example: the point $(-2, 5)$, reflected across the *y*-axis, becomes $(2, 5)$.]

- across the *x*-axis by changing $(x, y)$ to $(x, -y)$

    [Example: the point $(-4, 5)$, reflected across the *x*-axis, becomes $(-4, -5)$.]

- across the line $y = x$ by changing $(x, y)$ to $(y, x)$

    [Example: the point $(5, 7)$, reflected across the line $y = x$, becomes $(7, 5)$.]

- across the line $y = -x$ by changing $(x, y)$ to $(-y, -x)$

    [Example: the point $(-3, -4)$, reflected across the line $y = -x$, becomes $(4, 3)$.]

**Example:** A rectangle with coordinates (0, 3), (0, −2), (4, 3), (4, −2) is reflected across the *x*-axis. What are the coordinates of the reflected rectangle?

To produce a reflection across the *x*-axis, each *y*-coordinate must be replaced by its opposite. The new rectangle has coordinates (0, −3), (0, 2), (4, −3), (4, 2).

The examples of a translation, a rotation, and a reflection given above are for polygons, but the same principles apply to the simpler geometric elements of points and lines. In fact, a transformation performed on a polygon can be viewed equivalently as the same transformation performed on the set of points (vertices) and lines (sides) that compose the polygon. Thus, to perform complicated transformations on a figure, it is helpful to perform the transformations on all the points (or vertices) of the figure, then reconnect the points with lines as appropriate.

## 3.7 Multiple transformations

Multiple transformations can be performed on a geometrical figure. The order of these transformations may or may not be important. For instance, multiple translations can be performed in any order, as can multiple rotations (around a single fixed point) or reflections (across a single fixed line). The order of the transformations becomes important when several types of transformations are performed or when the point of rotation or the line of reflection changes among transformations. For example, consider a translation of a given distance upward and a clockwise rotation by 90° around a fixed point. Changing the order of these transformations changes the result.

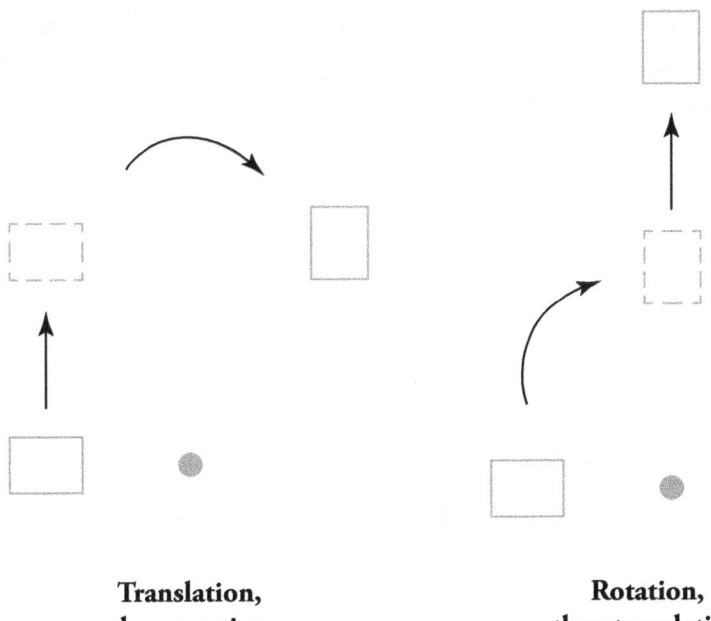

**Translation,**            **Rotation,**
**then rotation**          **then translation**

As shown, the final position of the box is different, depending on the order of the transformations. Thus, it is crucial that the proper order of transformations (whether determined by the details of the problem or some other consideration) be followed.

**Example:** Find the final location of a point at (1, 1) that undergoes the following transformations: rotate 90° counterclockwise about the origin; translate distance 2 in the negative *y*-direction; reflect about the *y*-axis.

First, draw a graph of the *x*- and *y*-axes and plot the point at (1, 1).

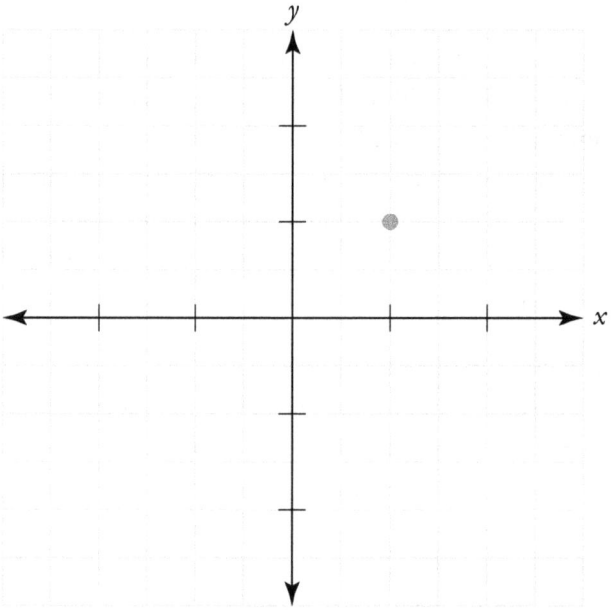

Next, perform the rotation. The center of rotation is the origin, and the rotation is in the counterclockwise direction. In this case, the even value of 90° makes the rotation simple to do by inspection. Next, perform a translation of distance 2 in the negative y direction (down). The results of these transformations are shown below.

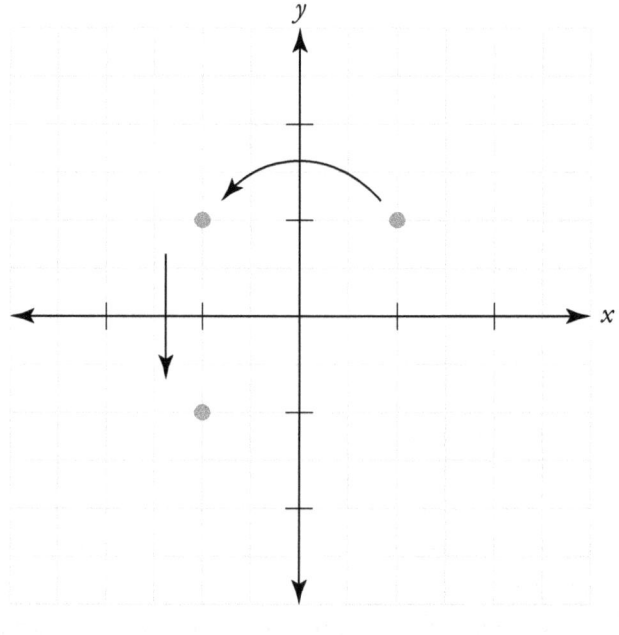

Finally, perform the reflection about the y-axis. The final result, shown below, is a point at $(1, -1)$.

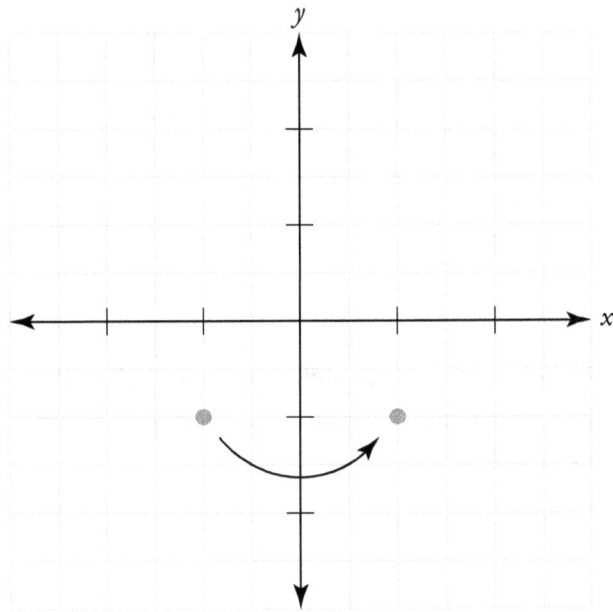

Using this approach, polygons can be transformed on a point-by-point basis. For some problems, there is no need to work with coordinate axes. For instance, the problem may simply require transformations without respect to any absolute positioning.

Example: Rotate the following regular pentagon by 36° about its center, and then reflect it about a horizontal line.

First, perform the rotation. In this case, the direction is not important because the pentagon is symmetric. As it turns out in this case, a rotation of 36° yields the same result as flipping the pentagon vertically (assuming the vertices of the pentagon are indistinguishable).

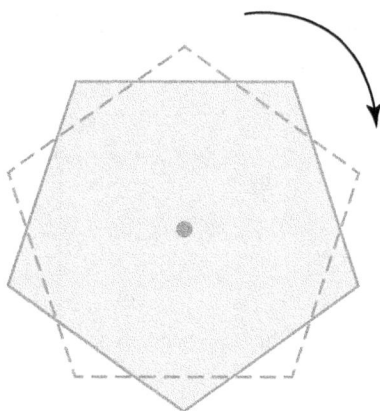

Finally, perform the reflection. Note that the result here is the same as a downward translation (assuming the vertices of the pentagon are indistinguishable).

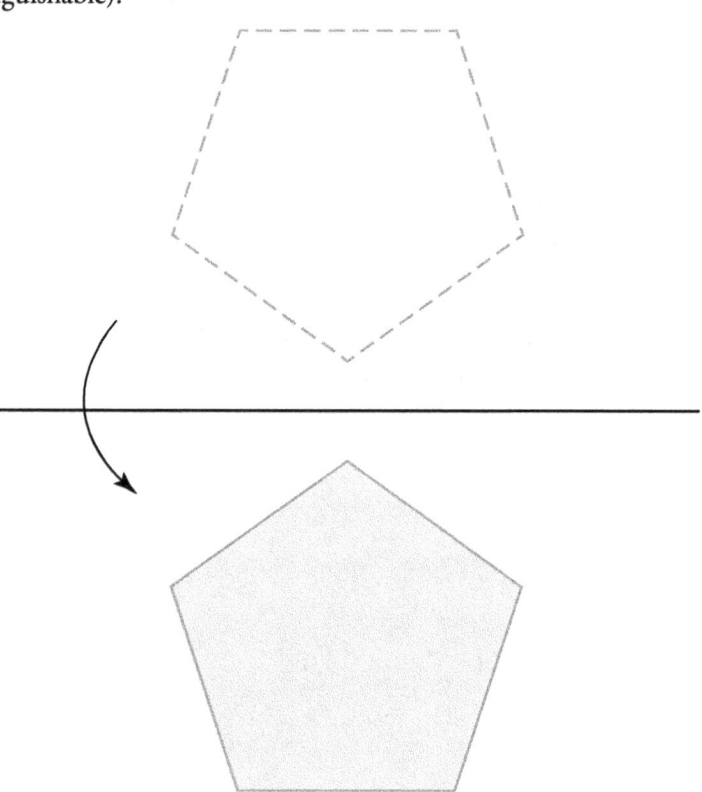

## 3.8  Symmetry

**Symmetry** is a property of a figure whose parts can be made to match up perfectly in a changed position. There are two kinds: reflectional symmetry and rotational symmetry.

### Reflectional symmetry

A figure has **reflectional symmetry** if there is a line, called a **line of symmetry**, that the figure could be folded across so that the halves would match completely. A figure can have more than one line of symmetry. A square, for instance, has four lines of symmetry, as the figure shows.

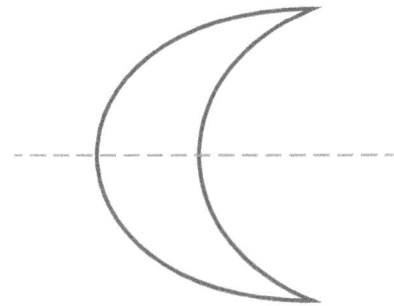
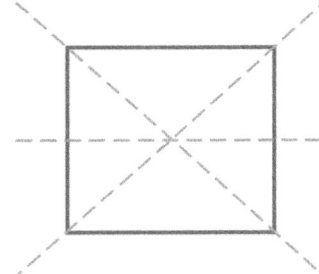

One line of symmetry    Four lines of symmetry

## Rotational symmetry

A figure has **rotational symmetry** if it would look the same after being rotated less than 360°. If the figure is unchanged by a rotation of 180°, its rotational symmetry can be described either as **order 2 rotational symmetry** (because the figure can appear in 2 identical positions), or as **180° rotational symmetry**. The recycling symbol below has rotational symmetry of order 3 or 120°.

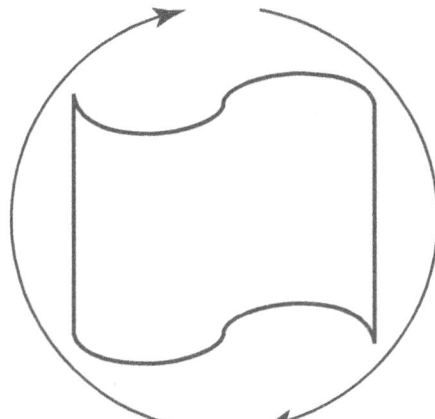
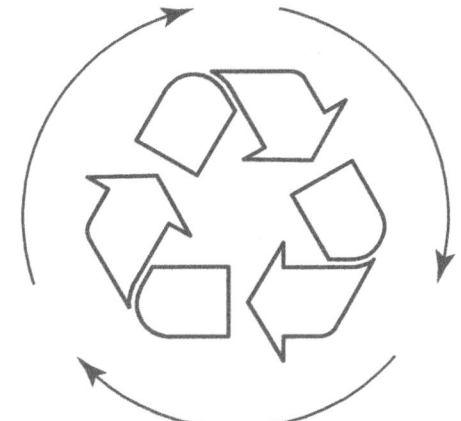

Order 2 symmetry    Order 3 symmetry

## Symmetry on the coordinate plane

Symmetries in a function can also be described in terms of reflections or "mirror images." A function can be symmetric about the $y$-axis (but not about the $x$-axis, except for the function $f(x) = 0$, since every function must pass the vertical line test). A function is symmetric about the $y$-axis if for every point $(x, y)$ that is included on the graph of the function, the point $(-x, y)$ is also included on the graph. Consider the function $f(x) = x^2$. Note that for each point $(x, x^2)$ on the graph, the point $(-x, x^2)$ is also on the graph. The symmetry of the function about the $y$-axis can also be seen in the graph below.

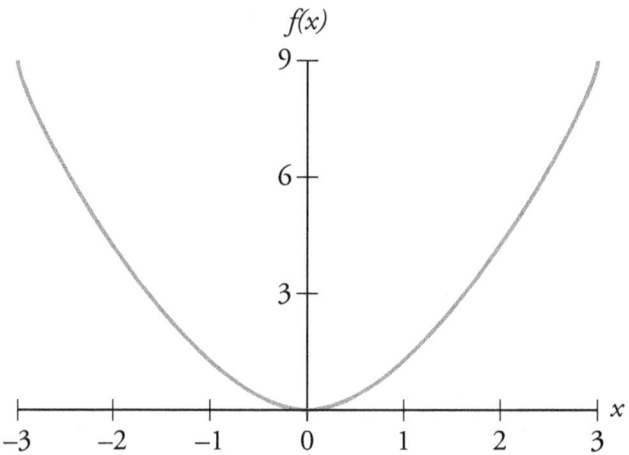

A function that is symmetric about the $y$-axis is also called an even function. Although functions cannot be symmetric about the $x$-axis, relations that do not obey the vertical line test can be symmetric in this way. A relation is symmetric about the $x$-axis if for every point $(x, y)$ in the graph of the relation, the point $(x, -y)$ is also in the graph.

Consider, for instance, the relation $g(x) = \pm\sqrt{x}$. For every value of $x$ in the domain, the points $(x, \sqrt{x})$ and $(x, -\sqrt{x})$ are both in the graph, as shown below.

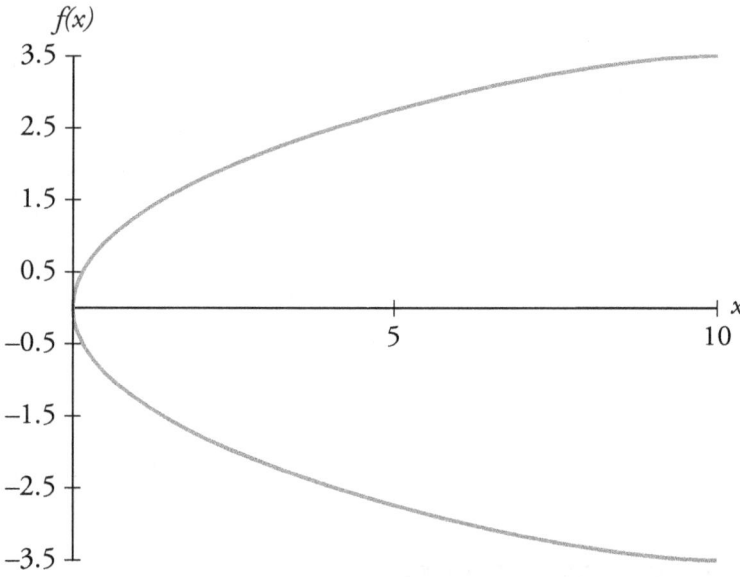

Functions may also be rotationally symmetric with respect to the origin. Such functions are called **odd (or antisymmetric) functions** and are defined by the property that for any point $(x, y)$ on the graph of the function, the point $(-x, -y)$ is also on the graph of the function. The function $f(x) = x^3$, for instance, is rotationally symmetric with respect to the origin, as shown in the graph below.

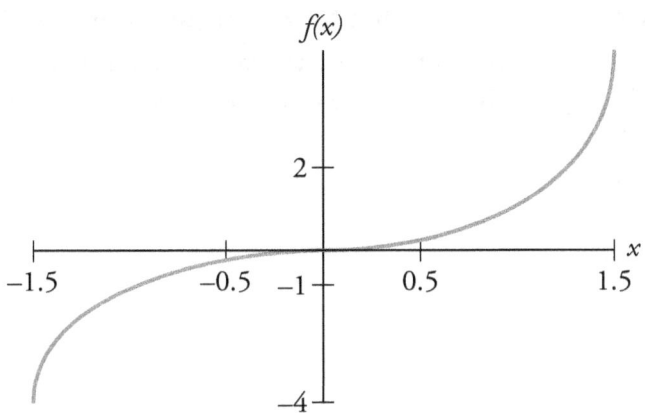

## 3.9 Polar coordinates

Any ordered pair can be plotted in a plane using either Cartesian (rectangular) or polar coordinates. **Cartesian coordinates** define a location by the horizontal distance from the y-axis and the vertical distance from the x-axis. **Polar coordinates** define a location as being a certain distance away from the origin along a path making a certain angle from the positive x-axis. In a pair of polar coordinates, the magnitude of the distance is the first coordinate and the angle of approach is the second coordinate.

These two representations are shown below for the example of (2, 3).

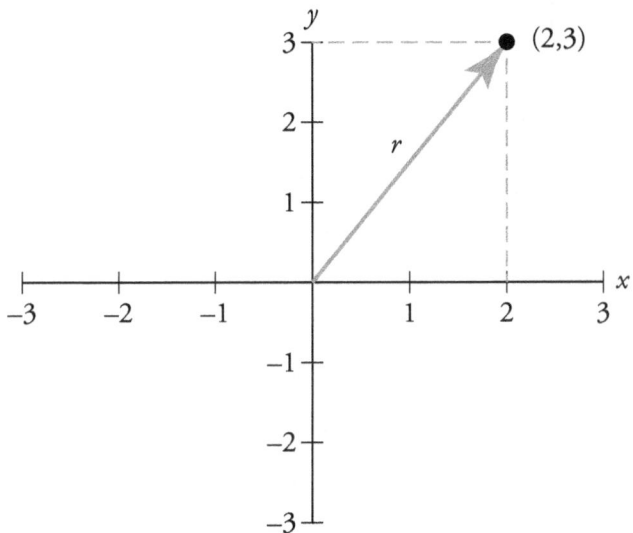

In the above case, r is $\sqrt{13}$ and $\theta$ is approximately 0.983 radians, so the point (2, 3) in Cartesian coordinates is $(\sqrt{13}, 0.983)$ in polar coordinates.

### Converting from rectangular coordinates to polar coordinates

Using the diagram above and the Pythagorean theorem, we can generalize as follows:

$r = \sqrt{x^2 + y^2}$

$\theta = \arctan\left(\dfrac{y}{x}\right)$

## Converting from polar coordinates to rectangular coordinates

To convert from polar coordinates to rectangular coordinates, apply right-triangle trigonometry to the diagram above for the general case of a point $(r, \theta)$ in polar coordinates.

$x = r \cos \theta$

$y = r \sin \theta$

# Chapter 4: Three-Dimensional Geometry

As two-dimensional figures have area and perimeter, three-dimensional figures have volume and surface area. The methods of finding volume and surface area vary with the type of figure.

## 4.1 Prisms

A **prism** is a three-dimensional figure whose two equal and parallel bases are polygons and whose sides are parallelograms. In a **right prism**, all the sides are rectangles. The volume and surface area of a prism can be derived by breaking the figure into portions for which these values can be calculated easily. For instance, consider the following **triangular prism**. Note that the two triangles are the bases of the prism even if they are oriented to front and back instead of being on top and on the bottom.

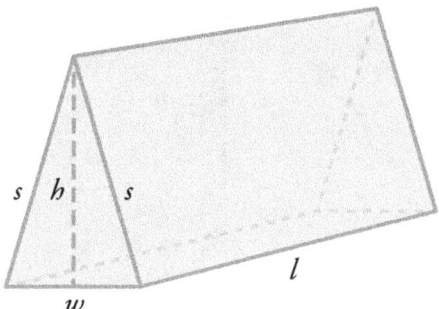

The volume of this figure can be found by calculating the area of the triangular cross section and then multiplying by the length $\ell$.

$$V = \frac{1}{2}hw\ell$$

The lateral surface area can be found by adding the areas of each side.

$$S = 2\left(\frac{1}{2}hw + s\ell\right) + \ell w$$

Similar reasoning applies to other figures composed of sides that are defined by triangles, quadrilaterals, and other planar or linear elements.

A **rectangular right prism** has two rectangular bases and four rectangular sides.
- Surface area $S = \ell w + hw + \ell h$, (where $\ell$ = length, $w$ = width, and $h$ = height)
- Volume $V = \ell w h$

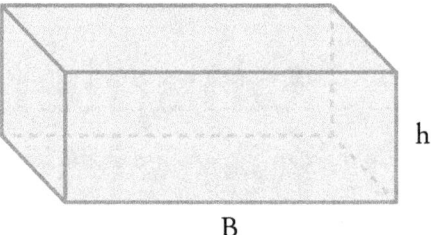

**Example:** Find the height of a box whose volume is 120 cubic meters if the area of the base is 30 square meters.

$V = Bh$
$120 = 30h$
$4 = h$

The height is 4 m.

## 4.2 Pyramids

A **pyramid** has a base that is a polygon and triangular sides meeting at a single point, the **vertex**. The surface area of a pyramid can be found by finding area of each side and adding to find the total surface area. The volume of a regular pyramid can be found with the following formula:

$V = \frac{1}{3}Bh$, where $B$ = area of the base of the pyramid and $h$ = the height of the pyramid

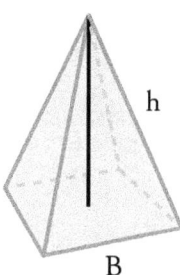

## 4.3 Curved figures

Below are formulas for the volumes and surface areas of the most important curved figures. A **cylinder** has two circular bases connected by a single curved side. Its volume and surface area are determined by the radius $r$ of its base and its height $h$. A **cone** has a circular base and a single curved side that comes to a point, the **vertex**. Its volume and surface area are also determined by the radius $r$ of its base and its height $h$. A **sphere** has one continuous curved surface consisting of points equidistant from a center point. Its volume and surface area are determined by its radius $r$. The formulas for the volume and surface area of these solids are given below.

| Figure | Volume | Surface Area |
|---|---|---|
| Cylinder | $V = \pi r^2 h$ | $SA = 2\pi rh + 2\pi r^2$ |
| Cone | $V = \dfrac{\pi r^2 h}{3}$ | $SA = \pi r\sqrt{r^2 + h^2} + \pi r^2$ |
| Sphere | $V = \dfrac{4}{3}\pi r^3$ | $SA = 4\pi r^2$ |

**Example:** How much material is needed to make a basketball that has a diameter of 15 inches? How much air is needed to fill the basketball?

Draw and label a sketch:

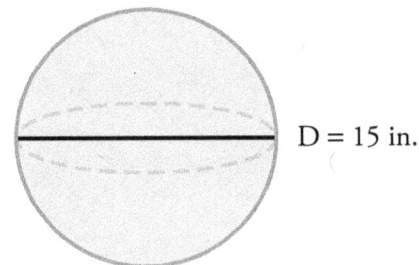

D = 15 in.

Surface Area     Volume

$SA = 4\pi r^2$     $V = \dfrac{4}{3}\pi r^3$     Write formula.

$= 4\pi(7.5)^2$     $= \dfrac{4}{3}\pi(7.5)^3$     Substitute 7.5 for $r$.

$= 706.9$ in.$^2$     $= 1767.1$ in.$^3$     Solve.

The basketball will require 706.9 in.² of material and will need to be filled with 1767.1 in.³ of air.

## 4.4   Similar solids and scale factors

**Similar solids** share the same shape but are not necessarily the same size. The ratio of any two corresponding linear measurements of similar solids is the **scale factor**. For example, the scale factor for two triangular prisms, one with a side measuring 2 inches and the other with a corresponding side measuring 4 inches, is 2:4.

The base perimeter, the surface area, and the volume of similar solids are all related to the scale factor, but not in the same way. If the scale factor of two similar solids is $a : b$, then
- the ratio of the base perimeters is also $a : b$
- the ratio of the areas is $a^2 : b^2$
- the ratio of the volumes is $a^3 : b^3$

Thus, for the above example,
- the ratio of the base perimeters is 2 : 4
- the ratio of the areas is 4 : 16
- the ratio of the volumes is 8 : 64

**Example:** What happens to the volume of a square pyramid when the lengths of the sides of the base and the height are doubled?

scale factor $a : b = 1 : 2$

ratio of volume $a^3 : b^3 = 1 : 8$

The volume is increased by a factor of 8.

**Example:** Given the following measurements for two similar cylinders with a scale factor of 2:5 (cylinder A to cylinder B), determine the height, radius, and volume of each cylinder.

cylinder A: $r = 2$

cylinder B: $h = 10$

For cylinder A,

$$\frac{h}{10} = \frac{2}{5}$$
$$5h = 20$$
$$h = 4$$

Volume of cylinder A $= \pi r^2 h = \pi(2^2)(4) = 16\pi$

For cylinder B,

$$\frac{2}{r} = \frac{2}{5}$$
$$2r = 10$$
$$r = 5$$

Volume of cylinder B $= \pi r^2 h = \pi(5^2)(10) = 250\pi$

## 4.5 Three-dimensional coordinates

To represent three-dimensional objects in a coordinate system, three coordinates are required. Thus, a point in three dimensions must be represented as $(x, y, z)$ instead of simply $(x, y)$, as in the two-dimensional representation. By convention, the $z$-coordinate represents height. So a point represented as $(3, 2, 5)$ is three units to the right of the origin, two units in front of the origin, and five units above the origin.

The distance between the origin and a point described by three-dimensional coordinates can be found using the Pythagorean theorem:

$$d = \sqrt{x^2 + y^2 + z^2}$$

**Example:** Find the distance between the origin and the point $(4, 12, 3)$.

$$d = \sqrt{x^2 + y^2 + z^2} = \sqrt{4^2 + 12^2 + 3^2} = \sqrt{169} = 13$$

The Pythagorean theorem can also be used to find the distance between two three-dimensional points by subtracting the first set of coordinates from the second to determine the x-distance, the y-distance and the z-distance between the two points. The formula in that case becomes

$$d = \sqrt{(x_2 - x_1)^2 + (y_2 - y_1)^2 + (z_2 - z_1)^2}$$

**Example:** Find the distance between the points $(2, -5, 0)$ and $(-5, -1, -4)$.

$$\begin{aligned} d &= \sqrt{(x_2 - x_1)^2 + (y_2 - y_1)^2 + (z_2 - z_1)^2} \\ &= \sqrt{(-5 - 2)^2 + (-1 - (-5))^2 + (-4 - 0)^2} \\ &= \sqrt{(-7)^2 + (4)^2 + (-4)^2} = \sqrt{81} = 9 \end{aligned}$$

# Chapter 5: Trigonometry

## 5.1   Trigonometric functions

**Trigonometric functions** can be related to right triangles: each trigonometric function corresponds to a ratio of certain sides of the triangle with respect to a particular angle. In the diagram below, the following functions can be specified.

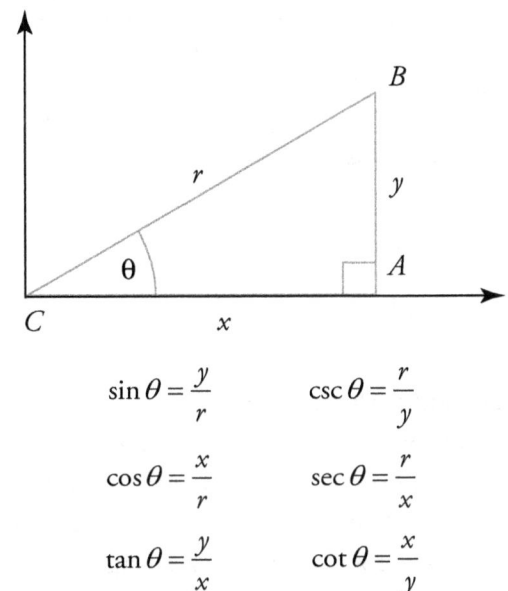

$$\sin\theta = \frac{y}{r} \qquad \csc\theta = \frac{r}{y}$$

$$\cos\theta = \frac{x}{r} \qquad \sec\theta = \frac{r}{x}$$

$$\tan\theta = \frac{y}{x} \qquad \cot\theta = \frac{x}{y}$$

Based on these definitions, the unknown characteristics of a particular right triangle can be calculated based on certain known characteristics. For instance, if the hypotenuse and one of the adjacent angles are both known, the lengths of the other two sides of the triangle can be calculated.

**Example:**   Find the length of the missing side of the triangle to the nearest tenth.

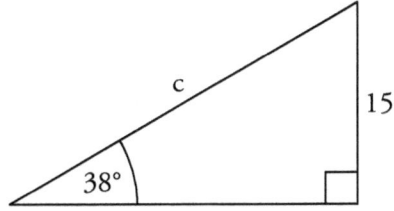

Based on the information provided in the diagram, the sine ratio should be used to solve for the hypotenuse, $c$.

$$\sin(38) = \frac{15}{c}$$
$$c\sin(38) = 15$$
$$c = \frac{15}{\sin(38)} \approx 24.4$$

**Example:** Find the measure of angle $A$ to the nearest degree.

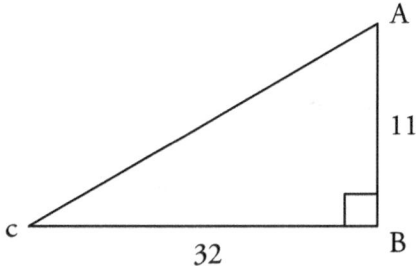

Based on the information provided in the diagram, the tangent ratio should be used to find the measure of angle $A$.

$$\tan(A) = \frac{32}{11}$$
$$A = \tan^{-1}\left(\frac{32}{11}\right) \approx 71°$$

## Angle measures: degrees and radians

The argument of a trigonometric function is an angle that is typically expressed in either degrees or radians. A **degree** constitutes an angle corresponding to a sector that is $\frac{1}{360}$ of a circle. Therefore, a circle has 360 degrees. A **radian**, on the other hand, is the angle corresponding to a sector of a circle whose arc length is equal to the radius of the circle (a little less than 60 degrees). In the case of the unit circle (a circle of radius 1), the circumference is $2\pi$. Thus, there are $2\pi$ radians in a circle.

Conversion between degrees and radians is a simple matter of using the ratio between the total degrees in a circle and the total radians in a circle.

$$\text{degrees} = \frac{180}{\pi} \times \text{radians}$$
$$\text{radians} = \frac{\pi}{180} \times \text{degrees}$$

## Trigonometry and the unit circle

Trigonometry can also be understood in terms of a unit circle on the *x-y* plane. A **unit circle** has a radius of 1.

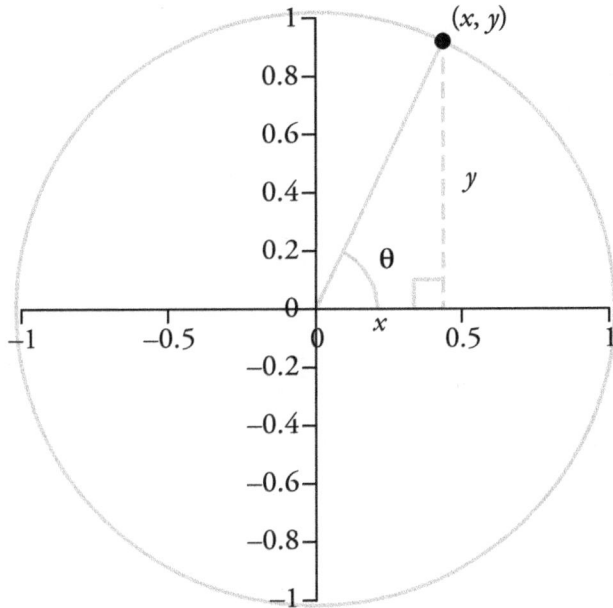

Notice that any given radius forms a right triangle with legs having lengths equal to the position of the point on the circle $(x, y)$. Since the radius is equal to 1, the values of $x$ and $y$ are the following:

$x = \cos \theta$

$y = \sin \theta$

All the properties of trigonometric relationships for right triangles also apply in this case as well.

## Graphs of trigonometric functions

The trigonometric functions sine, cosine, and tangent (and their reciprocals) are periodic functions. The values of **periodic functions** repeat at regular intervals.

The period, amplitude, and phase shift are critical properties of periodic functions that can be determined by observation of the graph or by detailed study of the functions themselves.

The **period** of a function is the smallest domain containing one complete cycle of the function. For example, the period of a sine or cosine function is the distance between the adjacent peaks or troughs of the graph. The **amplitude** of a function is half the distance between the maximum and minimum values of the function. The **phase shift** of a function is the amount of horizontal displacement of the function from a given reference position.

Below is a generic sinusoidal graph with the period and amplitude labeled.

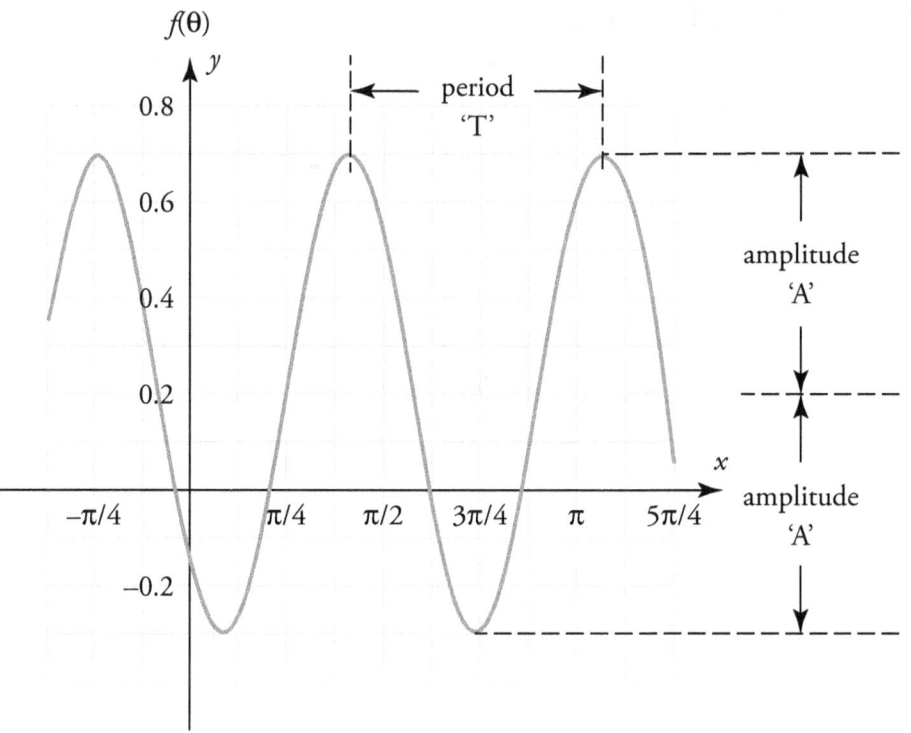

The period and amplitude for the three basic trigonometric functions are provided in the table below.

| Period and Amplitude of the Basic Trigonometric Functions | | |
|---|---|---|
| Function | Period (radians) | Amplitude |
| sin θ | $2\pi$ | 1 |
| cos θ | $2\pi$ | 1 |
| tan θ | $\pi$ | Undefined |

Below are the graphs of the basic trigonometric functions: (a) $y = \sin x$ (b) $y = \cos x$ (c) $y = \tan x$

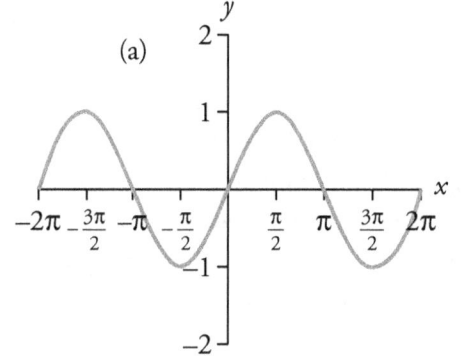

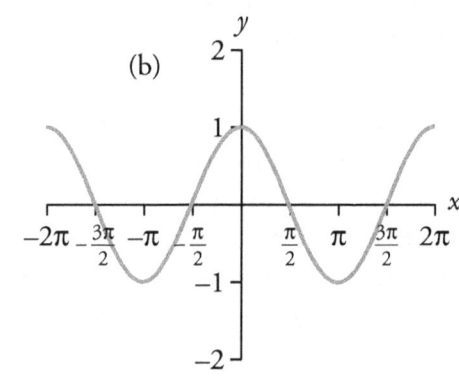

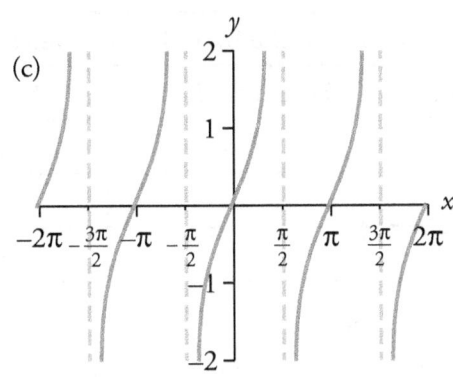
(c)

Note that the graph of the tangent function has asymptotes at $x = \frac{2n-1}{2}\pi$, where $n = 0, \pm 1, \pm 2, \pm 3 \ldots$

The reciprocal trigonometric functions are graphed below: (a) $y = \csc x$ (b) $y = \sec x$ (c) $y = \cot x$.

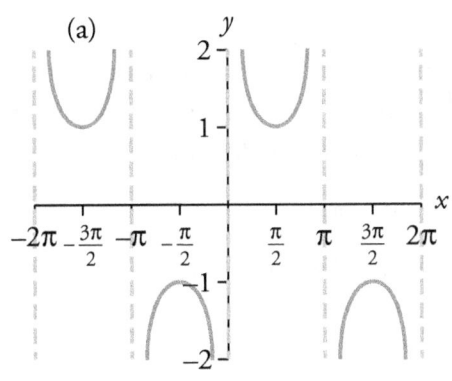
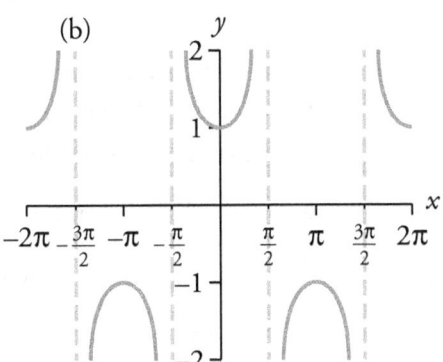
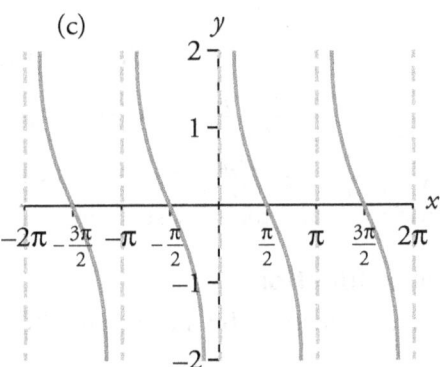

The phase and amplitude for the three reciprocal trigonometric functions are provided in the table below.

Chapter 5: Trigonometry    103

| Period and Amplitude of the Basic Trigonometric Functions | | |
| --- | --- | --- |
| Function | Period (radians) | Amplitude |
| csc θ | $2\pi$ | Undefined |
| sec θ | $2\pi$ | Undefined |
| cot θ | $\pi$ | Undefined |

## Solving trigonometric equations

The periodic nature of trigonometric functions means that trigonometric equations may have more than one solution within a given interval.

**Example:** Solve $2\cos x = \sqrt{3}$ for $x$ over the interval $[0, 360°]$

$$2\cos x = \sqrt{3}$$

First isolate the trigonometric function:
$$\cos x = \frac{\sqrt{3}}{2}$$

Then identify locations on the unit circle where the cosine has a value of $\frac{\sqrt{3}}{2}$. Cosine has a positive value in the first and fourth quadrants. The angles that correspond to this cosine ratio, then, are 30 and 330 degrees.

**Example:** Solve $\sin^2(3x) = 1$ for $x$ over the interval $[0, 2\pi]$

Take the plus or minus square root of both sides of the equation:

$$\sin^2(3x) = 1$$
$$\sin(3x) = \pm 1$$

Then identify locations on the unit circle where the sine has a value of $\pm 1$.

This occurs at $\frac{\pi}{2}, \frac{3\pi}{2}, \frac{5\pi}{2}, \frac{7\pi}{2}, \frac{9\pi}{2}...$. So $3x = \frac{\pi}{2}, \frac{3\pi}{2}, \frac{5\pi}{2}, \frac{7\pi}{2}, \frac{9\pi}{2}...$.

Then $x = \frac{\pi}{6}, \frac{\pi}{2}, \frac{5\pi}{6}, \frac{7\pi}{6}, \frac{3\pi}{2}, \frac{11\pi}{6}$ (any further extension of this pattern would represent solutions beyond the requested interval of $[0, 2\pi]$ )

## Graphing a trigonometric function

Graphing a trigonometric function by hand typically requires a calculator for determining the value of the function for various angles. Nevertheless, simple functions can often be graphed by simply determining the amplitude, period, and phase shift. Once these parameters are known, the graph can be sketched approximately. The amplitude of a simple sine or cosine function is simply the multiplicative constant (or function) associated with the trigonometric function. Thus, $y = 2\cos x$, for instance, has an amplitude of 2. The phase shift is typically just a constant added to the argument of the function. For instance, $y = \sin(x + 1)$ includes a phase shift of 1. A positive phase shift indicates that the graph of the function is shifted to the left; a negative phase shift indicates that the graph is shifted to the right.

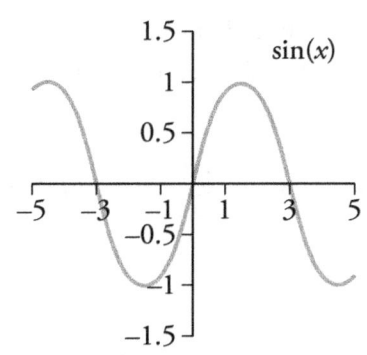
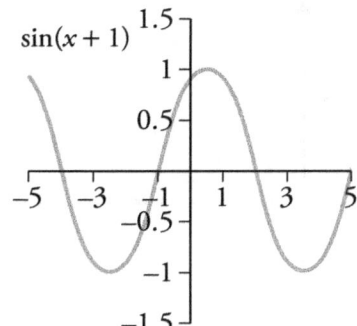
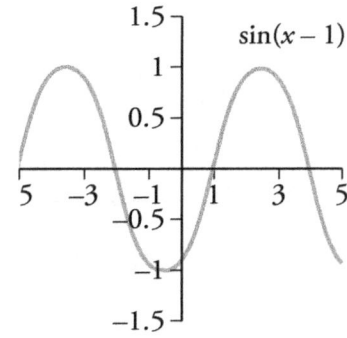

**Example:** Sketch the graph of the function $f(x) = 4\sin\left(2x + \dfrac{\pi}{2}\right)$

Notice first that the amplitude of the function is 4. Since there is no constant term added to the sine function, the function is centered on the x-axis. Find crucial points on the graph by setting $f$ equal to zero and solving for $x$ to find the roots.

$$f(x) = 0 = 4\sin\left(2x + \dfrac{\pi}{2}\right)$$

$$\sin\left(2x + \dfrac{\pi}{2}\right) = 0$$

$$2x + \dfrac{\pi}{2} = n\pi$$

In the above expression, $n$ is an integer.

$$2x = \left(n - \dfrac{1}{2}\right)\pi$$

$$x = \left(n - \dfrac{1}{2}\right)\dfrac{\pi}{2}$$

So, the roots of the function are at

$$x = \pm\dfrac{\pi}{4}, \pm\dfrac{3\pi}{4}, \dfrac{5\pi}{4}, \ldots$$

The maxima and minima of the function are halfway between successive roots. Determine the location of a maximum by testing the function. Try $x = 0$.

$$f(0) = 4\sin\left(2(0) + \frac{\pi}{2}\right) = 4\sin\left(\frac{\pi}{2}\right) = 4 \ f(0) 5 4 \sin(2[0] 1 p \cdot\cdot$$

Thus, $f$ is maximized at $x = 4$. The function can then be sketched.

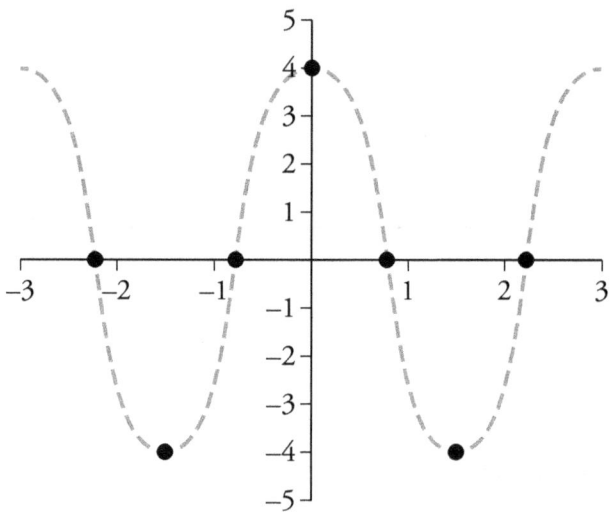

### Inverse trigonometric functions

The **inverse sine function** of $x$ is written as arcsin $x$ or $\sin^{-1} x$ and is the angle for which the sine is $x$; or, to put it another way, $\sin(\arcsin x) = x$. Since the sine function is periodic, many values of arcsin $x$ correspond to a particular $x$. In order to define arcsin as a function, therefore, its range needs to be restricted.

In some books, a restricted inverse function is denoted by a capitalized beginning letter such as in $\text{Sin}^{-1}$ or Arctan. The arcsin function is shown below.

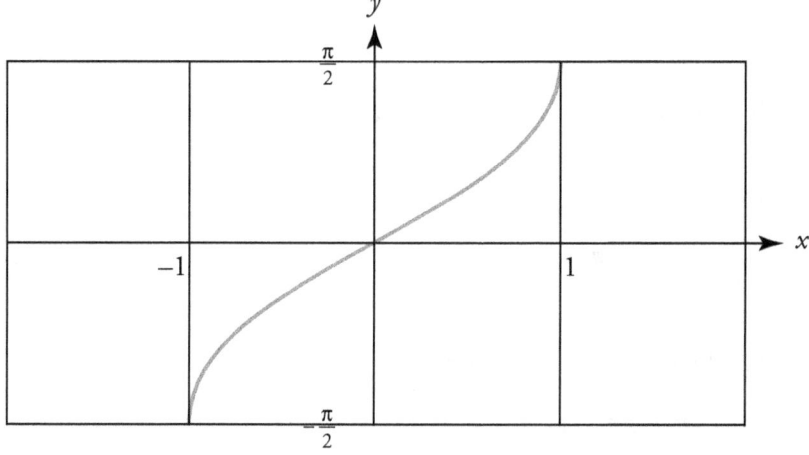

The inverse cosine and tangent functions are defined in the same way: cos(arccos x) = x; tan(arctan x) = x.

The function y = arccos x has a domain [−1,1] and range [0,π]. The graph of this function is shown below.

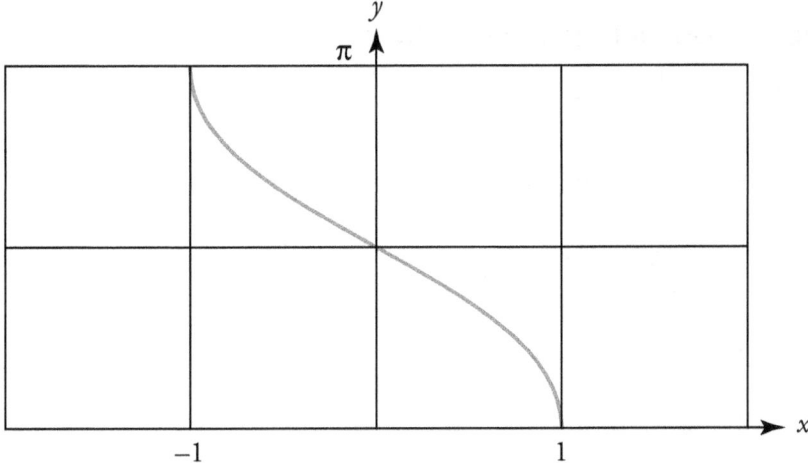

The function y = arctan x has a domain (−∞,+∞) and range $[-\frac{\pi}{2},\frac{\pi}{2}]$. The plot of the function is shown below.

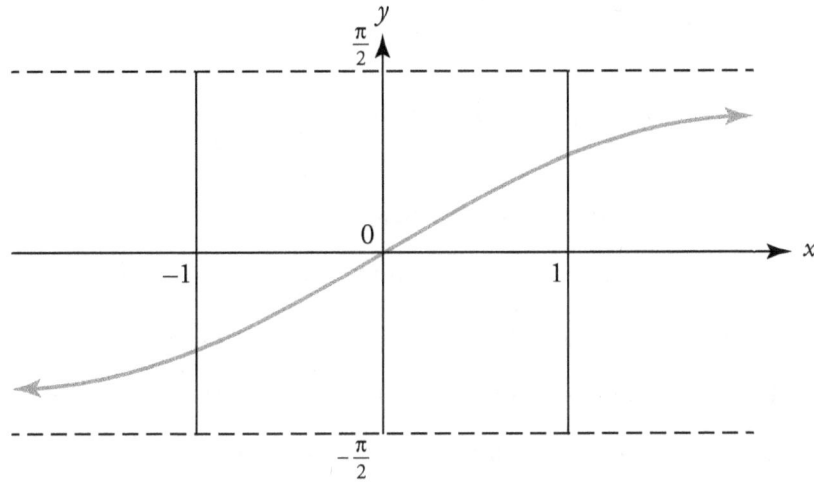

**Example:** Evaluate the following: (i) $\sin^{-1}(0)$; (ii) arccos(−1)

(i) $\sin(\sin^{-1}(0)) = 0$

The value of the inverse sine function must lie in the range $[-\frac{\pi}{2},\frac{\pi}{2}]$. Since 0 is the only argument in the range $[-\frac{\pi}{2},\frac{\pi}{2}]$ for which the sine function is zero, $\sin^{-1}(0) = 0$.

(ii) cos(arccos(−1)) = −1

The value of the inverse cosine function must lie in the range $[0,\pi]$ [0, p]. p is the only argument for which the cosine function is equal to {1 in the range [0, p].

Hence, $\arccos(-1) = \pi$.

## Identities for inverse trigonometric functions

$$\csc^{-1}(x) = \sin^{-1}\left(\frac{1}{x}\right) \text{ for } |x| \geq 1$$

$$\sec^{-1}(x) = \cos^{-1}\left(\frac{1}{x}\right) \text{ for } |x| \geq 1$$

$$\cot^{-1}(x) = \tan^{-1}\left(\frac{1}{x}\right) \text{ for } x > 0$$

$$= \tan^{-1}\left(\frac{1}{x}\right) + \pi \text{ for } x < 0$$

$$= \frac{\pi}{2} \text{ for } x = 0$$

$$\sin^{-1}(x) = \cos^{-1}\left(\sqrt{1-x^2}\right) \qquad \cos^{-1}(x) = \sin^{-1}\left(\sqrt{1-x^2}\right)$$

$$\tan^{-1}(x) = \cos^{-1}\left(\frac{1}{\sqrt{1+x^2}}\right) \qquad \cos^{-1}(x) = \tan^{-1}\left(\frac{\sqrt{1-x^2}}{x}\right)$$

$$\tan^{-1}(x) = \sin^{-1}\left(\frac{x}{\sqrt{1+x^2}}\right) \qquad \sin^{-1}(x) = \tan^{-1}\left(\frac{x}{\sqrt{1-x^2}}\right)$$

**Example:** Simplify the expression $\cos(\arcsin x) + \sin(\arccos x)$.

$\arcsin x = \arccos\left(\sqrt{1-x^2}\right)$ (identity) $\rightarrow \cos(\arcsin x) = \sqrt{1-x^2}$

$\arccos x = \arcsin\left(\sqrt{1-x^2}\right)$ (identity) $\rightarrow \sin(\arccos x) = \sqrt{1-x^2}$

Hence, $\cos(\arcsin x) + \sin(\arccos x) = \sqrt{1-x^2} + \sqrt{1-x^2} = 2\sqrt{1-x^2}$

**Example:** Using the identities given above, prove the identity $\sin^{-1} x + \cos^{-1} x = \frac{\pi}{2}$.

Since $\sin\left(\frac{\pi}{2}\right) = 1$, the identity may be proven by showing that

$\sin(\sin^{-1} x + \cos^{-1} x) = 1$

$\sin(\sin^{-1} x + \cos^{-1} x) = \sin(\sin^{-1} x)\cos(\cos^{-1} x) + \cos(\sin^{-1} x)\sin(\cos^{-1} x)$

sine sum formula

$$= x(x) + \sqrt{1-x^2}\left(\sqrt{1-x^2}\right)$$

$$= x^2 + 1 - x^2$$

$$= 1$$

Other similar identities include the following:

$$\tan^{-1} x + \cot^{-1} x = \frac{\pi}{2}$$

$$\sec^{-1} x + \csc^{-1} x = \frac{\pi}{2}$$

## 5.2 The Law of Sines

Trigonometric functions can also be applied to nonright triangles by way of the law of sines and the law of cosines. Consider the arbitrary triangle shown below with angles $A$, $B$ and $C$ and corresponding opposite sides $a$, $b$ and $c$.

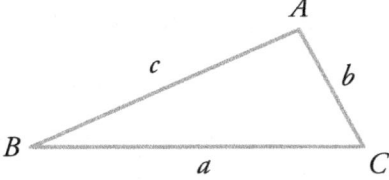

The **law of sines** is a proportional relationship between the lengths of the sides of a triangle and the opposite angles. The law of sines is given below:

$$\frac{a}{\sin A} = \frac{b}{\sin B} = \frac{c}{\sin C}$$

**Example:** An inlet is 140 feet wide. The lines of sight from each bank to an approaching ship are 79 degrees and 58 degrees. What are the distances from each bank to the ship?

First, draw an appropriate sketch of the situation with the appropriate labels for the parameters.

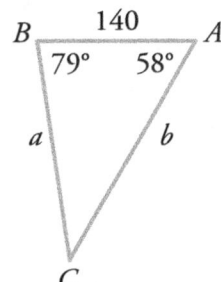

Since the sum of the angles in a triangle is 180°, angle $C$ must be 43°. Use the law of sines to calculate the lengths of sides $a$ and $b$.

For side $b$:

$$\frac{b}{\sin 79°} = \frac{140}{\sin 43°}$$

$$b = \left(\frac{\sin 79°}{\sin 43°}\right) 140 \approx 201.5 \text{ ft}$$

And for side $a$:

$$\frac{a}{\sin 58°} = \frac{140}{\sin 43°}$$

$$a = \frac{\sin 58°}{\sin 43°} 140 \approx 174.1 \text{ ft}$$

## 5.3 The law of cosines

The **law of cosines** permits determination of the length of a side of an arbitrary triangle as long as the lengths of the other two sides, along with the angle opposite the unknown side, are known. The law of cosines is given below:

$$c^2 = a^2 + b^2 - 2ab \cos C$$

**Example:** A triangular fence is being constructed around a water hazard. The first side is 96 m long. The second side, at a 65° angle to the first side, is 60 m long. How long must the third side be to complete the fence?

$$c^2 = a^2 + b^2 - 2ab \cos C$$
$$= 96^2 + 60^2 - 2(96)(60)\cos 65°$$
$$= 9216 + 3600 - 4868.5624$$
$$\approx 7947.4376$$
$$c \approx \sqrt{7947.4376} \approx 89.15$$

The third side will be about 89.15 m long.

## 5.4 Sum and difference formulas

Trigonometric functions involving the **sum or difference of two angles** can be expressed in terms of functions of each individual angle using the following formulas.

$$\sin(\alpha + \beta) = \sin\alpha \cos\beta + \cos\alpha \sin\beta$$
$$\cos(\alpha + \beta) = \cos\alpha \cos\beta - \sin\alpha \sin\beta$$
$$\tan(\alpha + \beta) = \frac{\tan\alpha + \tan\beta}{1 - \tan\alpha \tan\beta}$$
$$\sin(\alpha - \beta) = \sin\alpha \cos\beta - \cos\alpha \sin\beta$$
$$\cos(\alpha - \beta) = \cos\alpha \cos\beta + \sin\alpha \sin\beta$$
$$\tan(\alpha - \beta) = \frac{\tan\alpha - \tan\beta}{1 + \tan\alpha \tan\beta}$$

**Example:** Evaluate the following using the appropriate identity.

$$\sin(35°)\cos(55°) + \cos(35°)\sin(55°)$$

Using the sine sum formula,

$$\sin(35°)\cos(55°) + \cos(35°)\sin(55°) = \sin(35° + 55°) = \sin(90°) = 1$$

**Example:** Show that $\dfrac{\cos(x+y)}{\cos x \cos y} = 1 - \tan x \tan y$

Apply the cosine sum formula:

$$\dfrac{\cos(x+y)}{\cos x \cos y} = \dfrac{\cos x \cos y - \sin x \sin y}{\cos x \cos y}$$

$$= 1 - \dfrac{\sin x \sin y}{\cos x \cos y}$$

$$= 1 - \tan x \tan y$$

## 5.5 Double- and half-angle identities

The **double-angle identities** and **half-angle identities** are summarized below.

$\sin(2\alpha) = 2\sin\alpha\cos\alpha$

$\cos(2\alpha) = \cos^2\alpha - \sin^2\alpha = 1 - 2\sin^2\alpha$

$\tan(2\alpha) = \dfrac{2\tan\alpha}{1-\tan^2\alpha}$

$\sin\left(\dfrac{\alpha}{2}\right) = \pm\sqrt{\dfrac{1-\cos\alpha}{2}}$

$\cos\left(\dfrac{\alpha}{2}\right) = \pm\sqrt{\dfrac{1+\cos\alpha}{2}}$

$\tan\left(\dfrac{\alpha}{2}\right) = \pm\sqrt{\dfrac{1-\cos\alpha}{1+\cos\alpha}}$ or $\dfrac{\sin\alpha}{1+\cos\alpha}$ or $\dfrac{1-\cos\alpha}{\sin\alpha}$

The double-angle identities can be obtained using the sum formulas, as with the example of the cosine function below:

$\cos(2\alpha) = \cos(\alpha + \alpha)$

$= \cos\alpha\cos\alpha - \sin\alpha\sin\alpha$ cos 2 a 5 cos (a 1 a)

$= \cos^2\alpha - \sin^2\alpha$

**Example:** Show that $\sin(3x) = \sin x(3\cos^2 x - \sin^2 x)$.

$\sin(3x) = \sin(2x + x)$

$\quad = \sin 2x \cos x + \cos 2x \sin x$ Sine sum formula

$\quad = 2\sin x\cos x\cos x + (\cos^2 x - \sin^2 x)\sin x$ Double-angle formula

$\quad = 2\sin x\cos^2 x + \sin x\cos^2 x - \sin^3 x$

$\quad = 3\sin x\cos^2 x - \sin^3 x$

$\quad = \sin x(3\cos^2 x - \sin^2 x)$

The half-angle identities can be derived by solving the double angle identities for the sine, cosine, or tangent of a single angle. For instance,

$$\cos(2\alpha) = 2\cos^2 a - 1$$
$$2\cos^2 a = 1 + \cos(2\alpha)$$
$$\cos^2 a = \frac{1+\cos(2\alpha)}{2}$$
$$\cos a = \sqrt{\frac{1+\cos(2\alpha)}{2}}$$

Since this identity is valid for all values of $a$, it will continue to be valid if we replace $a$ with $\frac{\alpha}{2}$. Therefore,

$$\cos\frac{a}{2} = \pm\sqrt{\frac{1+\cos\alpha}{2}}$$

Note that the choice of the appropriate sign depends on the value of the angle $\frac{\alpha}{2}$. If $\frac{\alpha}{2}$ is in the first or fourth quadrants, then the positive sign is chosen. Otherwise, the negative sign must be chosen. The choice of the sign is in conformity with the standard characteristics of trigonometric functions.

**Example:** Given that $\sin 30° = \frac{1}{2}$, find the value of $\tan 15°$.

$$\cos 30° = \sqrt{1-\sin^2 30°} = \sqrt{1-\frac{1}{4}} = \sqrt{\frac{3}{4}} = \frac{\sqrt{3}}{2}$$

$$\tan 15° = \frac{1-\cos 30°}{\sin 30°} \quad \text{Half-angle identity}$$

$$= \frac{1-\frac{\sqrt{3}}{2}}{\frac{1}{2}} = 2-\sqrt{3} \approx 0.27$$

## 5.6 Proving trigonometric identities

There are two methods that may be used to prove **trigonometric identities**. One method is to choose one side of the equation and manipulate it until it equals the other side. The other method is to replace expressions on both sides of the equation with equivalent expressions until both sides are equal.

There are a range of trigonometric identities, including **reciprocal** and **Pythagorean identities**, as listed below.

### Reciprocal identities

$\sin x = \dfrac{1}{\csc x}$  $\qquad$  $\sin x \csc x = 1$  $\qquad$  $\csc x = \dfrac{1}{\sin x}$

$\cos x = \dfrac{1}{\sec x}$  $\qquad$  $\cos x \sec x = 1$  $\qquad$  $\sec x = \dfrac{1}{\cos x}$

$\tan x = \dfrac{1}{\cot x}$  $\qquad$  $\tan x \cot x = 1$  $\qquad$  $\cot x = \dfrac{1}{\tan x}$

$\tan x = \dfrac{\sin x}{\cos x}$  $\qquad\qquad\qquad\qquad$  $\cot x = \dfrac{\cos x}{\sin x}$

**Pythagorean identities**

$$\sin^2 x + \cos^2 x = 1 \qquad 1 + \tan^2 x = \sec^2 x \qquad 1 + \cot^2 x = \csc^2 x$$

**Example:** Prove that $\sin^2 x + \cos^2 x = 1$.

Use the definitions of the sine and cosine functions from right-triangle trigonometry.

$$\left(\frac{y}{r}\right)^2 + \left(\frac{x}{r}\right)^2 = 1$$

$$\frac{x^2 + y^2}{r^2} = 1$$

$$x^2 + y^2 = r^2$$

The identity resolves into a statement of the Pythagorean theorem, which is already proven. For this reason it is called a Pythagorean identity.

**Example:** Prove that $\cot x + \tan x = \csc x \sec x$

Use the reciprocal identities to convert the left side of the equation to sines and cosines. Then combine terms using a common denominator.

$$\frac{\cos x}{\sin x} + \frac{\sin x}{\cos x} = \frac{\cos^2 x}{\sin x \cos x} + \frac{\sin^2 x}{\sin x \cos x}$$

$$= \frac{\sin^2 x + \cos^2 x}{\sin x \cos x}$$

$$= \frac{1}{\sin x \cos x} \qquad \text{Pythagorean identity}$$

$$= \csc x \sec x \qquad \text{Reciprocal identities}$$

# Chapter 6: Data Analysis, Statistics and Probability

## 6.1 Factorials and their uses

The **factorial** of a positive integer is the product of that integer and every lesser positive integer down to 1. The factorial of 5, for instance, which is written as 5!, equals $5 \times 4 \times 3 \times 2 \times 1$. By convention, the factorial of 0 is 1.

**Permutations and combinations**

A **permutation** is one of the possible arrangements of $n$ items, without repetition, where the order of selection is important.

A **combination** is one of the possible arrangements of $n$ items, without repetition, where the order of selection is not important.

**Example:** If any two numbers are selected from the set {1, 2, 3, 4}, list the possible permutations and combinations.

| Combinations | Permutations |
|---|---|
| 12, 13, 14, 23, 24, 34 | 12, 21, 13, 31, 14, 41, 23, 32, 24, 42, 34, 43 |
| six ways | twelve ways |

Note that the list of permutations includes 12 and 21 as separate possibilities, since the order of selection is important. In the case of combinations, however, the order of selection is not important and, therefore, 12 is the same combination as 21. Hence, 21 is not listed separately as a possibility.

The number of permutations and combinations may also be found by using the formulae given below.

The number of possible permutations in selecting $r$ objects from a set of $n$ objects is given by

$$_nP_r = \frac{n!}{(n-r)!}$$

The notation $_nP_r$ is read "the number of permutations of $n$ objects taken $r$ at a time." In our example, two objects are being selected from a set of four.

$$_4P_2 = \frac{4!}{(4-2)!} = \frac{4 \times 3 \times 2 \times 1}{2 \times 1} = \frac{24}{2} = 12$$

The number of possible combinations in selecting $r$ objects from a set of $n$ objects is given by $_nC_r = \dfrac{n!}{(n-r)!r!}$

In our example,

$$_4C_2 = \dfrac{4!}{(4-2)!2!} = \dfrac{24}{2(2)} = 6$$

### Objects arranged in a row

It can be shown that $_nP_n$, the number of ways $n$ objects can be arranged in a row, is equal to $n!$. We can imagine $n$ positions being filled, one at a time. The first position can be filled in $n$ ways using any one of the $n$ objects.

Since one of the objects has been used, the second position can be filled in only $(n-1)$ ways. Similarly, the third position can be filled in $(n-2)$ ways, and so on. Hence, the total number of possible arrangements of $n$ objects in a row is given by

$$_nP_n = n(n-1)(n-2)\ldots 1 = n!$$

**Example:** Five books are placed in a row on a bookshelf. In how many different ways can they be arranged?

The number of possible ways in which 5 books can be arranged in a row is

$5! = 5 \times 4 \times 3 \times 2 \times 1 = 120$

The formula given above for $_nP_r$, the number of possible permutations of $r$ objects selected from $n$ objects, can also be proved in a similar manner. If $r$ positions are filled by selecting from $n$ objects, the first position can be filled in $n$ ways, the second position can be filled in $n-1$ ways, and so on (as shown before). The $r$th position can be filled in $n-(r-1) = n-r+1$ ways. Hence,

$$_nP_r = n(n-1)(n-2)\ldots(n-r+1) = \dfrac{n!}{(n-r)!}$$

The formula for the number of possible combinations of $r$ objects selected from $n$ objects, $_nC_r$, may be derived by using the above two formulae. For the same set of $r$ objects, the number of permutations is $r!$. All these permutations, however, correspond to the same combination. Hence,

$$_nC_r = \dfrac{_nP_r}{r!} = \dfrac{n!}{(n-r)!r!}$$

## 6.2  Measures of central tendency and dispersion in a dataset

It is often desirable to offer some very brief summary of what a set of data is like. Various ways of summarizing a dataset are available, depending on what quality of the data you wish to show. **Measures of central tendency** show in different ways where the center or the heart of a dataset lies or what sort of data item would be typical. The mean, median, and mode are measures of central tendency (i.e., the average or typical value) in a data set. They can be defined both for discrete and continuous data sets. **Measures of dispersion** show how far the data diverge from the center. Range, variance, and standard deviation are measures of dispersion.

## Mean

For discrete data, the **mean** is the average of the data items, or the value obtained by adding all the data values and dividing by the total number of data items.

Example: Find the mean of the following temperatures in degrees Fahrenheit:

71, 82, 65, 93, 87, 79, 82

There are 7 items, so the mean is

$$\frac{71+82+65+93+88+79+82}{7} = \frac{560}{7} = 80$$

## Weighted averages

Sometimes it is desirable to assign more importance to some items in a dataset than to others. That can be done by finding a **weighted average**. Any item in the dataset can be multiplied by a weighting factor k as it is added in to the numerator. That item is also counted, not just as 1 item, but as $k$ items, when tallying the denominator.

Example: Your average in a class is determined by your scores on three quizzes and a final exam. The final exam counts double. Your scores on the quizzes and the final exam, are 83, 91, 86, and 90. What is your average?

Since the final exam counts double, that score is doubled when adding up the total of the scores. That test also counts as 2 tests in the denominator, so the total is divided by 5 instead of 4.

Your average is $\dfrac{83+91+86+2(90)}{1+1+1+2} = \dfrac{440}{5} = 88$

## Median

The **median** is found by putting the data in order from smallest to largest and selecting the value in the middle (or the average of the two values in the middle if the number of data items is even).

Example: The ages of the members of the school board are 66, 42, 54, 49, 67, and 58. What is the median age?

From smallest to largest, the items are 42, 49, 54, 58, 66, and 67. Since there are 6 items, an even number, the median is halfway between the two middle items, 54 and 58, so the median is 56.

## Mode

The **mode** is the most frequently occurring data value. There can be more than one mode in a data set, if there is a tie between two or more values for frequency of occurrence.

**Example:** What is the mode of the following dataset: 22, 35, 22, 37, 25, 26, 31, 35, 29

The most frequently appearing values are 22 and 35, tied at two appearances apiece. Both 22 and 35 are the mode.

### Range

The **range** is a measure of variability that is calculated by subtracting the smallest value from the largest value in a dataset.

**Example:** What is the range of the following data?

42, 35, 57, 28, 87, 63, 54

The smallest value is 28. The largest value is 87. The range is $87 - 28 = 59$.

### Interquartile range

**Interquartile range** is a measure of central tendency that is used in the construction of box plots. To find the interquartile range of a dataset, first find the median. Then separate the data into two groups, the data below the median and the data above. If the dataset has an odd number of items, the median will fall on one of the items, which will be excluded from either group since it is neither to the left nor the right of itself. If there is an even number of items, the median will fall between two items. In that case, all the data items will be in one group or the other. Now find the median of the lower group, which is called the first quartile. Find the median of the upper group, which is called the third quartile. The difference between the two quartiles is the interquartile range.

**Example:** Find the interquartile range of the following data:

7, 10, 11, 12, 13, 16, 19, 21, 24, 35

There are 10 items, an even number. The median is halfway between the fifth and sixth items, $\frac{13+16}{2} = 14.5$. The median of the 5 items of the left is 11, so that is the first quartile. The third quartile, or median of the upper group, is 21. Finally, the interquartile range is $21 - 11 = 10$.

### Variance and standard deviation

**Variance** and **standard deviation** are measures of how widely or narrowly the data are spread around the mean. A low variance or standard deviation suggests that the data are clustered around a center. A high variance or standard deviation shows that the data area spread out more widely.

To calculate the variance and standard deviation:
1. Calculate the mean by adding all the data items and dividing by the number of items.
2. Find the difference between each data item and the mean.
3. Find the square of each difference.
4. Add up all the squared differences.

5. Divide the total of the squared differences by the number of data items. That is the variance.
6. Take the square root of the variance. That is the standard deviation.

Example: Calculate the variance and standard deviation for the following data set:

{3, 3, 5, 7, 8, 8, 8, 10, 12, 21}.

The mean is $\frac{3+3+5+7+8+8+8+10+12+21}{10} = \frac{85}{10} = 8.5$

The difference of each item $x$ from the mean is $|8.5 - x|$. The 10 differences are:

5.5, 5.5, 3.5, 1.5, 0.5, 0.5, 0.5, 1.5, 3.5, 12.5

The squares of those ten differences are

30.25, 30.25, 12.25, 2.25, 0.25, 0.25, 0.25, 2.25, 12.25, 156.25

The sum of the ten squared differences is

30.25 + 30.25 + 12.25 + 2.25 + 0.25 + 0.25 + 0.25 + 2.25 + 12.25 + 156.25
= 246.5

The mean of the ten squared differences is

$\frac{30.25+30.25+12.25+2.25+0.25+0.25+0.25+2.25+12.25+156.25}{10}$

$= \frac{246.5}{10} = 24.65$

The variance, written as $\sigma^2$, is 24.65.

The standard deviation $\sigma$ is the square root of the variance. $\sigma = \sqrt{24.65} \approx 4.96$

## 6.3 Displaying statistical data

The data obtained from sampling may be categorical (e.g., yes or no responses) or numerical. In both cases, results are displayed using a variety of graphical techniques. Geographical data is often displayed superimposed on maps.

### Histograms

The most common form of graphical display used for numerical data obtained from random sampling is the histogram. A **histogram** displays data by dividing the data values into equal-sized ranges, called **bins**, and displaying bars whose height indicates the number of data items that fall into that bin. A trend line or curve can be superposed on a histogram to observe the general shape of the distribution.

If the data set is large, it may be expressed in compact form as a frequency distribution. The number of occurrences of each data point is the frequency of that value. The relative frequency is defined as the frequency divided by the total number of data points. Since the sum of the frequencies equals the number of data points, the relative frequencies add up to 1. The relative frequency of a data point, therefore, represents the probability of occurrence of that value. Thus, a distribution consisting of relative frequencies is known as a probability distribution.

The cumulative frequency of a data point is the sum of the frequencies from the beginning up to that point. A histogram is used to display a discrete frequency distribution graphically. It shows the counts of data in different ranges, the center of the data set, the spread of the data, and whether there are any outliers. It also shows whether the data has a single mode or more than one.

**Example:** The table below shows the summary of some test results, where people scored points ranging from 0 to 45. The total range of points has been divided into bins 0–5, 6–10, 11–15, and so on. The frequency for the first bin (labeled 5) is the number of people who scored points ranging from 0 to 5; the frequency for the second bin (labeled 10) is the number of people who scored points ranging from 6 to 10; and so on.

| Points | Frequency | Cumulative Frequency | Relative Frequency |
|---|---|---|---|
| 5 | 1 | 1 | 0.009 |
| 10 | 4 | 5 | 0.035 |
| 15 | 12 | 17 | 0.105 |
| 20 | 22 | 39 | 0.193 |
| 25 | 30 | 69 | 0.263 |
| 30 | 25 | 94 | 0.219 |
| 35 | 13 | 107 | 0.114 |
| 40 | 6 | 113 | 0.053 |
| 45 | 1 | 114 | 0.009 |

The histogram of the probability distribution is given below:

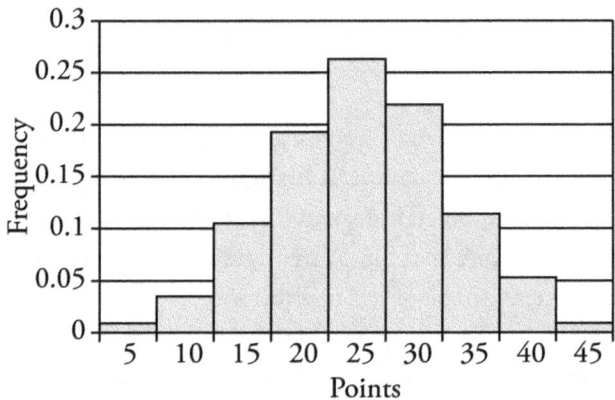

The probability distribution can be used to calculate the probability of a particular test score occurring in a certain range. For instance, the probability of a test score lying between 15 and 30 is given by the sum of the areas (assuming width of 1) of the three middle bins in the histogram above:

0.193 +1 0.263 + 0.219 = 0.675

## Bar graphs

**Bar graphs** are used to compare various quantities using bars of different lengths.

Example: A class had the following grades: 4 A's, 9 B's, 8 C's, 1 D, 3 F's.

Graph these on a bar graph.

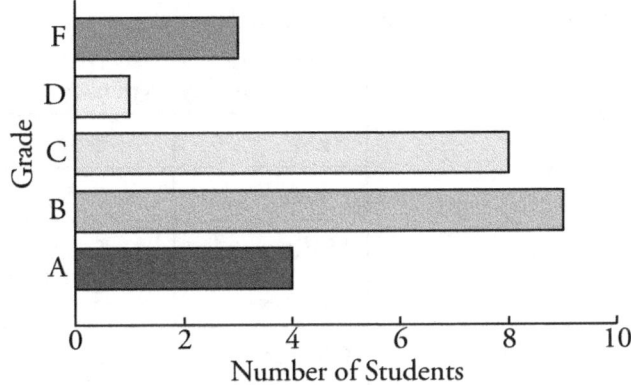

## Line graphs

**Line graphs** are used to show trends, often over a period of time.

Example: Graph the following information using a line graph.

| Number of National Merit Finalists/School Year | | | | | | |
|---|---|---|---|---|---|---|
| School | 90–91 | 91–92 | 92–93 | 93–94 | 94–95 | 95–96 |
| Central | 3 | 5 | 1 | 4 | 6 | 8 |
| Wilson | 4 | 2 | 3 | 2 | 3 | 2 |

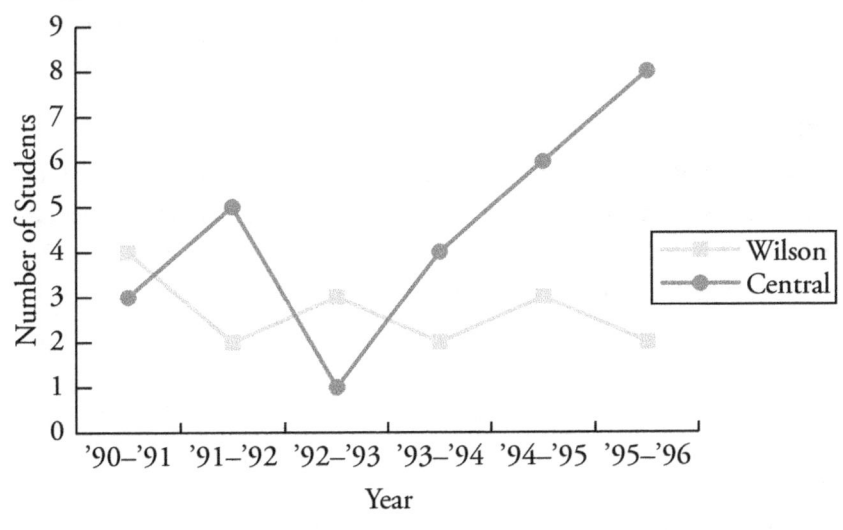

### Circle graphs (pie charts)

**Circle graphs** or **pie charts** show the relationships of various parts of a data set to each other and to the whole. Each part is shown as a percentage of the total and occupies a proportional sector of the circular area. To make a circle graph, total all the information that is to be included on the graph. Determine the central angle to be used for each sector of the graph using the following formula:

% of whole × 360° = central angle of sector

Lay out the respective central angles of the various sectors, label each section and include its percentage.

**Example:** Graph this information on a circle graph:

| Monthly Expenses | |
| --- | --- |
| Rent | $400 |
| Food | $150 |
| Utilities | $ 75 |
| Clothes | $ 75 |
| Church | $100 |
| Misc. | $200 |

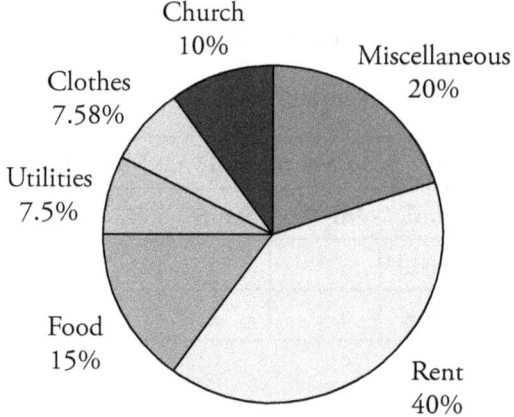

### Scatter plots

**Scatter plots** compare two characteristics of the same group of things or people and usually consist of a large body of data. They show how much one variable is affected by another. The relationship between the two variables is their correlation. The closer the data points come to making a straight line when plotted, the closer the correlation.

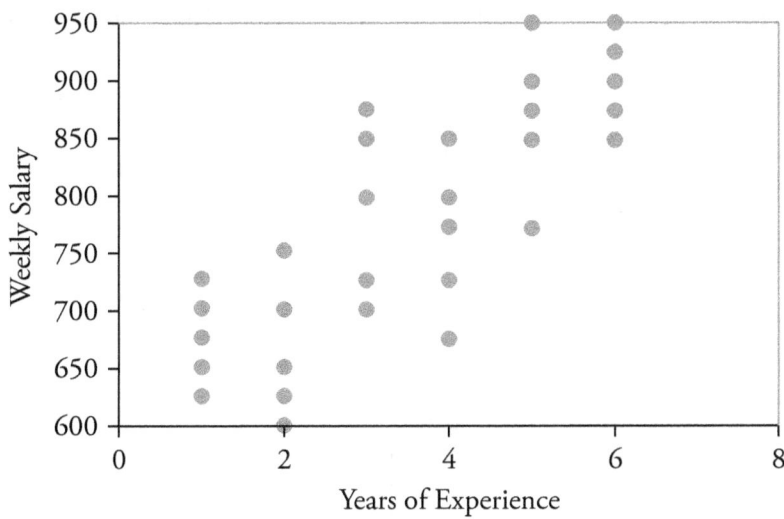

## 6.4 Regression models

It is often helpful to use regression to construct a more general trend or distribution based on sample data. **Regression** is a process of using the data to construct a modeling function that will approximately represent those data. To select an appropriate model for the regression, a representative set of data must be examined. It is often helpful, in this case, to plot the data and review it visually on a graph. In this manner, it is relatively simple to select a general class of functions (linear, quadratic, exponential, etc.) that might be used to model the data. There are two basic aspects of regression: selection of an appropriate curve that best fits the data and quantification of the closeness of fit of that curve. For instance, if a line can be constructed that passes through every data point of a distribution, then that line is a perfect fit to the data (and, obviously, linear regression is an appropriate choice for the model). If the distribution of data points seems to bear no particular resemblance to the line, then linear regression is probably not a wise choice, and a quantification of the closeness of fit should reflect this fact.

The following discussion summarizes least-squares linear regression analysis. The same principles can be applied to other forms of regression (such as quadratic or exponential).

### The method of least squares

Given a set of data, a curve approximation can be fitted to the data by using the **method of least squares**. The best-fit curve, defined by the function $f(x)$, is assumed to approximate a set of data with coordinates $(x_i, y_i)$ by minimizing the sum of squared differences between the curve and the data. Mathematically, the sum of these squared differences (errors) can be written as follows for a dataset with $n$ points.

$$S = \sum_{i=1}^{n} [f(x_i) - y_i]^2$$

Thus, the best-fit curve approximation to a set of data $(x_i, y_i)$ is an $f(x)$ that minimizes $S$. Shown below is a set of data and a linear function that approximates it. The vertical distances between the data points and the line are the errors that are squared and summed to find S.

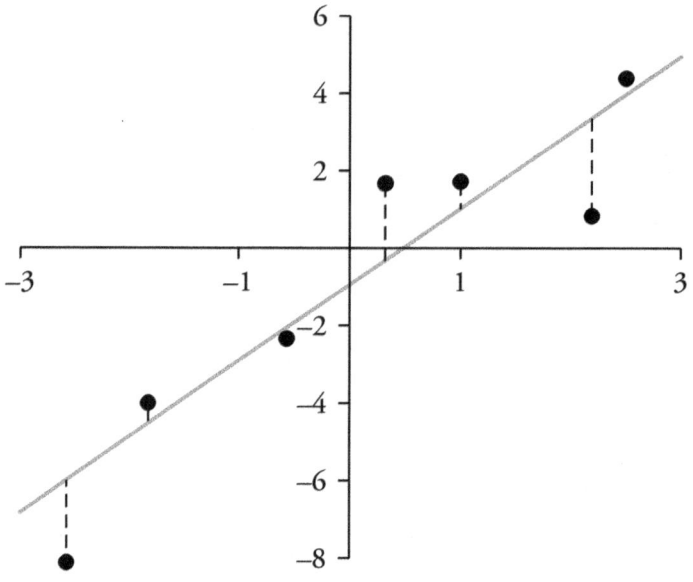

### Linear least-squares regression

If the function $f(x)$ that is used to approximate a set of data by minimizing the sum $S$ of squared errors (or residuals is linear, then $f(x)$ is called a **least-squares regression line**. The process of determining $f(x)$ is called linear least-squares regression. In this case, $f(x)$ has the form $f(x) = ax + b$.

Given a set of data $\{(x_1, y_1), (x_2, y_2), (x_3, y_3), \ldots (x_n, y_n)\}$, the sum $S$ for linear regression is the following.

$$S = \sum_{i=1}^{n}(ax_i + b - y_i)^2$$

To find $f(x)$, it is necessary to find $a$ and $b$. This can be done by minimizing $S$. Since $S$ is a function of both $a$ and $b$, $S$ must be minimized through the use of partial derivatives. (A partial derivative is exactly the same as a full derivative, except that all variables other than the one being differentiated are treated as constants. Partial derivatives often use the symbol $\partial$ in place of $d$.) Therefore, find the partial derivative with respect to $a$ and the partial derivative with respect to $b$.

$$\frac{\partial S}{\partial a} = \frac{\partial}{\partial a}\sum_{i=1}^{n}[ax_i + b - y_i]^2 \qquad \frac{\partial S}{\partial b} = \frac{\partial}{\partial b}\sum_{i=1}^{n}[ax_i + b - y_i]^2$$

$$\frac{\partial S}{\partial a} = \sum_{i=1}^{n}2x_i[ax_i + b - y_i] \qquad \frac{\partial S}{\partial b} = \sum_{i=1}^{n}2[ax_i + b - y_i]$$

Set these equal to zero. This yields a system of equations that can be solved to find $a$ and $b$. Although the algebra is somewhat involved, it is not conceptually difficult. The results are given below.

$$a = \frac{n\sum_{i=1}^{n} x_i y_i - \sum_{i=1}^{n} x_i \sum_{i=1}^{n} y_i}{n\sum_{i=1}^{n} x_i^2 - [\sum_{i=1}^{n} x_i]^2}$$

Note that the average *x*-value for the data (which is the sum of all *x*-values divided by *n*) and the average *y*-value for the data (which is the sum of all *y*-values divided by *n*) can be used to simplify the expression. The average *x*-value is defined as $\bar{x}$ and the average *y*-value is defined as $\bar{y}$.

$$a = \frac{\sum_{i=1}^{n} x_i y_i - n\bar{x}\bar{y}}{\sum_{i=1}^{n} x_i^2 - n\bar{x}^2}$$

Since the expression for *b* is complicated, it suffices to show *b* in terms of the above expression for *a*.

$$b = \frac{1}{n}\left(\sum_{i=1}^{n} y_i - a\sum_{i=1}^{n} x_i\right)$$

$$b = \bar{y} - a\bar{x}$$

Thus, given a set of data, the linear least-squares regression line can be found by calculating *a* and *b* as shown above.

## Correlation coefficient

The **correlation coefficient** *r* can be used as a measure of the quality of *f(x)* as a fit to the data set. The value of *r* ranges from zero (for a poor fit) to one (for a good fit). The correlation coefficient formula is given below.

$$r^2 = \frac{[\sum_{i=1}^{n} x_i y_i - \frac{1}{n}\sum_{i=1}^{n} x_i \sum_{i=1}^{n} y_i]^2}{[\sum_{i=1}^{n} x_i^2 - \frac{1}{n}(\sum_{i=1}^{n} x_i)^2][\sum_{i=1}^{n} y_i^2 - \frac{1}{n}(\sum_{i=1}^{n} y_i)^2]}$$

$$r^2 = \frac{(\sum_{i=1}^{n} x_i y_i - n\bar{x}\bar{y})^2}{(\sum_{i=1}^{n} x_i^2 - n\bar{x}^2)^2(\sum_{i=1}^{n} y_i^2 - n\bar{y}^2)}$$

**Example:** A company has collected data comparing the ages of its employees to their respective incomes (in thousands of dollars). Find the line that best fits the data (using a least-squares approach). Also calculate the correlation coefficient for the fit. The data is given below in the form of (age, income).

{(35, 42), (27, 23), (54, 43), (58, 64), (39, 51), (31, 40)}

The data are plotted in the graph below.

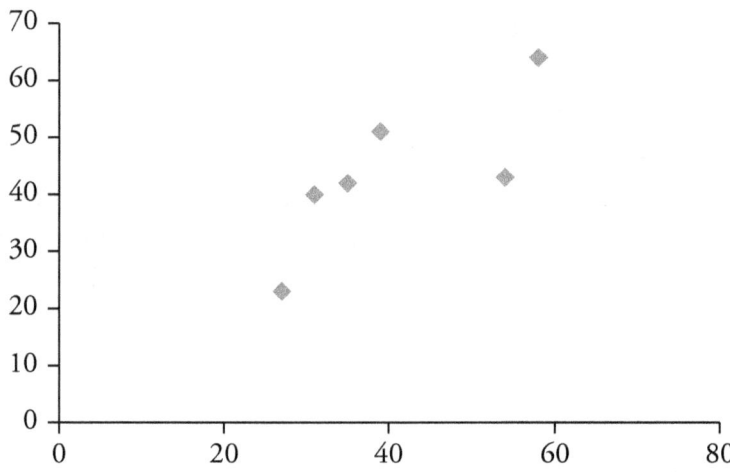

Note that there are six pieces of data. It is helpful to first calculate the following sums:

$$\sum_{i=1}^{6} x_i = 35 + 27 + 54 + 58 + 39 + 31 = 244$$

$$\sum_{i=1}^{6} y_i = 42 + 23 + 43 + 64 + 51 + 40 = 263$$

$$\sum_{i=1}^{6} x_i y_i = 35(42 + 27(23) + 54(43) + 58(64) + 39(51) + 31(40) = 11354\ 6o$$

$$\sum_{i=1}^{6} x_i^2 = 35^2 + 27^2 + 54^2 + 58^2 + 39^2 + 31^2 = 10716$$

$$\sum_{i=1}^{6} y_i^2 = 42^2 + 23^2 + 43^2 + 64^2 + 51^2 + 40^2 = 12439$$

Based on these values, the average $x$-and $y$-values are given below.

$$\bar{x} = \frac{244}{6} \approx 40.67$$
$$\bar{y} = \frac{263}{6} \approx 43.83$$

To find the equation of the least-squares regression line, calculate the values of $a$ and $b$.

Thus, the equation of the least-squares regression line is

$f(x) = 0.832x + 9.993$

This result can be displayed on the data graph to ensure that there are no egregious errors in the result.

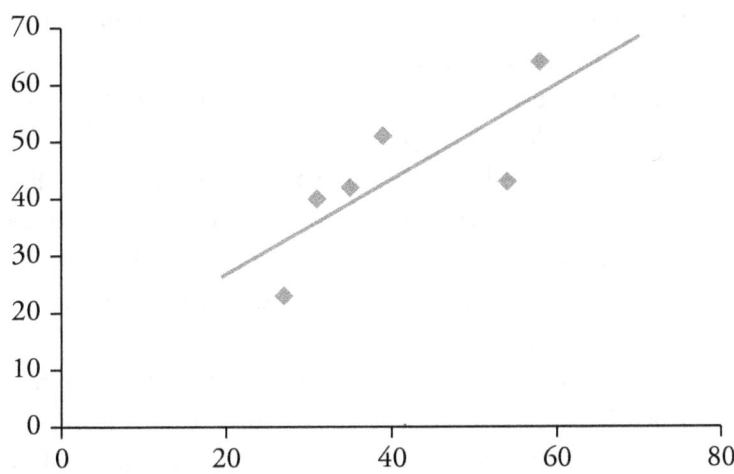

The regression line in the graph above appears to do a good job of approximating the trend of the data. To quantify how well the line fits the data, calculate the correlation coefficient using the formula given above.

$$r^2 = \frac{(11354 - 6(40.67)(43.83))^2}{(10716 - 6(40.67)^2)(12439 - 6(43.83)^2)}$$

$$r^2 = \frac{(658.603)^2}{(791.707)(912.587)} = 0.600$$

$$r \approx 0.775$$

Thus, the fit to the data is reasonably good.

### Calculators and regression

Modern handheld calculators, especially graphing calculators, often have built-in tools for handling regression. After entering the data (in the form of lists, for instance), the calculator's regression functions can be employed. It is usually best to test several different functions (if no particular model is obviously appropriate) and to compare the associated residual or correlation coefficient values for each function. For instance, it may be best to test a data set using both linear and exponential models. By comparing the correlation coefficient, the better-fitting curve can be determined. In this case, the closer the value is to unity, the better the fit. Of course, it is necessary to know the features of a particular calculator, as well as its limitations, to employ regression functions accurately.

## 6.5    Probability

The **probability** of an outcome, given a random experiment (a structured, repeatable experiment for which the outcome cannot be predicted or, alternatively, for which the outcome is dependent on "chance"), is the relative frequency of the outcome. The relative frequency of an outcome is the fraction or percentage of times an experiment yields that outcome for a very large (ideally, infinite) number of trials. For instance, if a fair coin is tossed a very large number of times, then the relative frequency of tossing heads is 0.5, or 50% (that is, one out of every two tosses, on average, should come out heads up). The probability is this relative frequency.

In probability theory, the **sample space** is a list of all possible outcomes of an experiment. For example, the sample space of tossing two coins is the set {HH, HT, TT, TH}, where H is heads and T is tails, and the sample space of rolling a six-sided die is the set {1, 2, 3, 4, 5, 6}. When conducting experiments with a large number of possible outcomes, it is important to determine the size of the sample space. The size of the sample space can be determined by using the fundamental counting principles and the rules of combinations and permutations.

A **random variable** is a function that corresponds to the outcome of some experiment or event, which is in turn dependent on "chance." For instance, the result of a tossed coin is a random variable: the outcome is either heads or tails, and each outcome has an associated probability. A **discrete variable** is one that can only take on certain specific values. For instance, the number of students in a class can only be a whole number (e.g., 15 or 16, but not 15.5). A **continuous variable**, such as the weight of an object, can take on a continuous range of values.

The probabilities for the possible values of a random variable constitute the probability distribution for that random variable. Probability distributions can be discrete, as with the case of the tossing of a coin (there are only two possible distinct outcomes), or they can be continuous, as with the outside temperature at a given time of day. In the latter case, the probability is represented as a continuous function over a range of possible temperatures, and finite probabilities can only be measured in terms of ranges of temperatures rather than specific temperatures. That is to say, for a continuous distribution, it is not meaningful to say "the probability that the outcome is $x$"; instead, only "the probability that the outcome is between $x$ and $x + k$" is meaningful. (Note that if each potential outcome in a continuous distribution has a nonzero probability, then the sum of all the probabilities would be greater than 1, since there are an infinite number of potential outcomes.)

Example: Find the sample space and construct a probability distribution for the sum of the outcomes from tossing two six-sided dice (each with numbers 1 through 6).

The sample space is the set of all possible outcomes that can arise in a given trial. For two dice, the outcome of any toss can be written as a two-digit number, with each digit being from 1 to 6. Using this convention, the sample space for tossing two 6-sided dice is as follows:

{11, 12, 13, 14, 15, 16,
 21, 22, 23, 24, 25, 26,
 31, 32, 33, 34, 35, 36,
 41, 42, 43, 44, 45, 46,
 51, 52, 53, 54, 55, 56,
 61, 62, 63, 64, 65, 66}

Notice that all the numbers in any diagonal rising from left to right have the same sum. We can show the frequency of the different sums in a histogram:

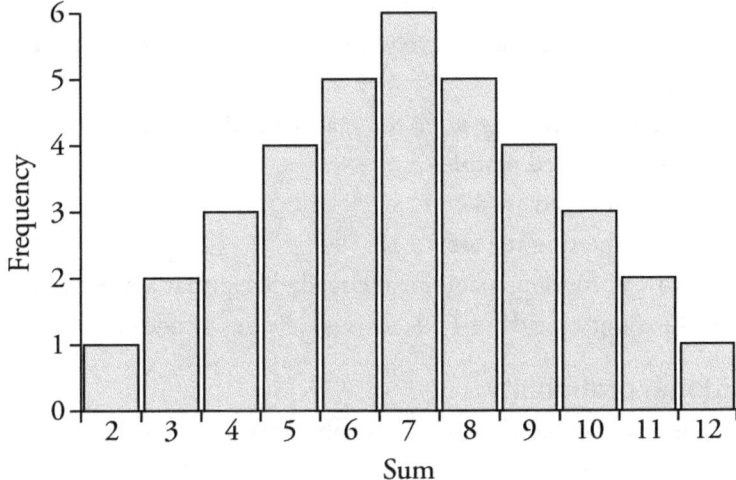

To construct the associated probability distribution, note first that the sum of the probabilities must equal 1. Since there are 36 possibilities in all, the probability of each sum occurring is its frequency divided by 36. Since the sum 6 occurs 5 times in the frequency distribution, the probability of rolling two dice that add to 6 is 5 out of 36 or $\frac{5}{36}$, or about 0.14.

The probability distribution can be shown as a histogram below.

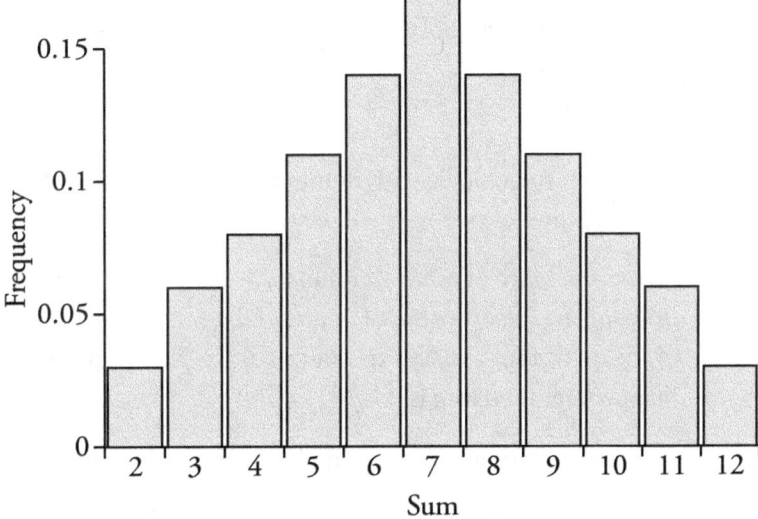

The sum of the probabilities for all the possible outcomes of a discrete distribution must be equal to unity. The expected value of a probability distribution is the same as the mean value of a probability distribution. The expected value is thus a measure of the central tendency or average value for a random variable with a given probability distribution.

### Odds

Probability can also be expressed in terms of odds. **Odds** are defined as the ratio of the number of favorable outcomes to the number of unfavorable outcomes. The sum of the favorable outcomes and the unfavorable outcomes should always equal the total possible outcomes.

For example, given a bag of 12 red marbles and 7 green marbles, compute the odds of randomly selecting a red marble.

Odds of getting red are 12 : 7.

Odds of not getting red are 7 : 12.

In the case of flipping a coin, it is equally likely that a head or a tail will be tossed, so the odds of tossing a head are 1 : 1. This is called even odds.

### Conditional probability

**Conditional probability** is the probability that a second event will happen, given that a first event has happened.

Example: Consider the following two events: the home team wins the semifinal round (event $A$) and the home team wins the final round (event $B$). The probability of event $B$ is contingent on the probability of event $A$. If the home team fails to win the semifinal round, it has a zero probability of winning in the final round. On the other hand, if the home team wins the semifinal round, then it may have a finite probability of winning in the final round. Symbolically, the probability of event $B$ given event $A$ is written $P(B|A)$.

Conditional probability can be calculated according to the following definition (the symbol $\cap$ means "and," the symbol $\cup$ < means "or," and $P(x)$ means "the probability of $x$"):

$$P(B\,|\,A) = \frac{P(A \cap B)}{P(A)}$$

Example: If the first roll of a die with numbers from 1 to 6 is a 3, what is the probability that the total of two rolls will be 6 or greater?

The probability of event $A$, rolling a 3, is 1 in 6 or $\frac{1}{6}$. For event $A$ and $B$ both to happen, the outcome of $A$ must be a 3 and the outcome of $B$ must be 3, 4, 5, or 6, four possibilities out of 6 or $\frac{4}{6}$. The probability of both events happening, therefore is $\frac{1}{6}\left(\frac{4}{6}\right) = \frac{4}{36} = \frac{1}{9}$.

The probability of $B$ happening in the event that $A$ happens, then, is

$$P(B\,|\,A) = \frac{P(A \cap B)}{P(A)} = \frac{1}{9} \div \frac{1}{6} = \frac{1}{9} \times \frac{6}{1} = \frac{6}{9} = \frac{2}{3}$$

### Probability of independent events

Two events are **independent** if the probability of the second event does not depend on the outcome of the first event. The probability that two independent events will both happen can be found by multiplying the separate probabilities:

$$P(A \cap B) = P(A)P(B)$$

**Example:** Consider a pair of dice: one red and one green. First the red die is rolled, followed by the green die. What is the probability of rolling a 2 with the red die and a 5 with the green die?

It is apparent that these events do not depend on each other, since the outcome of the roll of the green die is not affected by the outcome of the roll of the red die. Thus the events are independent events.

$$P(A \cap B) = P(A)P(B) = \frac{1}{6}\left(\frac{1}{6}\right) = \frac{1}{36}$$

## Probability of dependent events

Often events are not independent. Suppose a jar contains 12 red marbles and 8 blue marbles. If a marble is selected at random and then replaced, the probability of picking a certain color is the same in the second trial as it is in the first trial. If the marble is not replaced, then the probability of picking a certain color is not the same in the second trial, because the total number of marbles is decreased by 1. In this case, the second event is **dependent** on the first, because the first event influences the possibilities of the second event. This is an illustration of conditional probability. If $R_1$ is the probability of picking a red marble the first time and $R_2$ is the probability of picking a red marble the second time, then, *if the first marble is replaced before picking the second marble*, the probability of picking a red marble both times is simply the product of the separate probabilities for each of the two trials:

$$P(R_1 \cap R_2) = P(R_1)P(R_2) = \frac{12}{20}\left(\frac{12}{20}\right) = \frac{144}{400} = 0.36$$

If the first marble is *not replaced* before picking the second marble, and a red marble is chosen the first time, in the second trial there will be only 11 red marbles and 19 marbles to choose from. Therefore, the probability of picking a red marble both times equals the probability of picking a red marble in the first trial times the probability of picking a red marble in the second trial, assuming that a red marble has been picked in the first trial.

$$P(R_1 \cap R_2) = P(R_1)P(R_2 \mid R_1) = \frac{12}{20}\left(\frac{11}{19}\right) = \frac{132}{380} \approx 0.347$$

**Example:** A car has a 75% probability of traveling 20,000 miles without breaking down. It has a 50% probability of traveling 10,000 additional miles without breaking down if it first makes it to 20,000 miles without breaking down. What is the probability that the car reaches 30,000 miles without breaking down?

Let event $A$ be that the car reaches 20,000 miles without breaking down.

$P(A) = 0.75$

Chapter 6: Data Analysis, Statistics and Probability

Event *B* is that the car travels an additional 10,000 miles without breaking down (assuming it didn't break down for the first 20,000 miles). Since event *B* is contingent on event *A*, write the probability as follows:

$P(B|A) = 0.50$

Use the probability formula for dependent events to find the probability of $A \cap B$, the probability that the car will travel 30,000 miles without breaking down.

$P(A \cap B) = P(A)P(B|A) = 0.75(0.5) = 0.375$

The car has a 37.5% probability of traveling 30,000 miles without breaking down.

# SECTION III:
# SAT Math Level 2
# Practice Test 1

# SAT Math Level 2 Practice Test 1

## Sample Test Questions

*Instructions:* For each problem, select the choice that best answers the question.

1. Simplify $(2 - 3i)(4i)$.
   A. $-12 + 8i$
   B. $12 + 8i$
   C. $8 - 12i$
   D. $8 + 12i$
   E. $-8 + 12i$

2. Solve $\sqrt{n^2 + 16} = 3n$
   A. $2$
   B. $\pm 2$
   C. $\pm\sqrt{2}$
   D. $\pm\frac{4}{3}$
   E. No Real Solution

3. Solve for x such that $0 \leq x \leq 2\pi$: $\frac{1}{2}\sin(2x) - \frac{\sqrt{2}}{4} = 0$
   A. $\left\{\frac{\pi}{4}\right\}$
   B. $\left\{\frac{\pi}{8}\right\}$
   C. $\left\{\frac{\pi}{8}, \frac{\pi}{4}\right\}$
   D. $\left\{\pm\frac{\pi}{8}, \pm\frac{\pi}{4}\right\}$
   E. $\left\{\frac{\pi}{8}, \frac{3\pi}{8}, \frac{9\pi}{8}, \frac{11\pi}{8}\right\}$

4. Find the solution to the system of equations. $\begin{cases} 4x + 2y = 18 \\ y = -2x + 9 \end{cases}$
   A. No solution
   B. Infinitely many solutions
   C. $(2, 1)$
   D. $(9, 18)$
   E. $(0, 0)$

5. If $h(x) = \dfrac{3x+4}{x}$ for all real values of x ≠ 0, find $h^{-1}(x)$.

   A. $h^{-1}(x) = \dfrac{-3x-4}{x}$

   B. $h^{-1}(x) = \dfrac{x}{3x+4}$

   C. $h^{-1}(x) = \dfrac{4}{x-3}$

   D. $h^{-1}(x) = \left(\dfrac{1}{x}\right)(3x+4)$

   E. None of the above represent $h^{-1}(x)$.

6. What values of x will ensure that all values of this function are real numbers? $f(x) = \sqrt{3-2x}$

   A. $\left\{x \mid x \leq \dfrac{3}{2}\right\}$

   B. $\left\{x \mid x \geq \dfrac{2}{3}\right\}$

   C. $\left\{x \mid x - \dfrac{3}{2} \leq x \leq \dfrac{3}{2}\right\}$

   D. $\left\{x \mid x \neq \dfrac{3}{2}\right\}$

   E. All Real numbers, x

7. Which of the following would produce an odd number?

   A. $(Odd + Even + Odd) \times Odd$
   B. $(Odd \times Odd) + Odd$
   C. $(Even + Odd + Even) \times Odd$
   D. $(Odd + Odd) \times (Even + Odd)$
   E. $(Odd \times Odd) - Odd$

8. **Based on the given table, if $y_1 = x^3$, what is the equation for $y_2$?**

| X | −2 | −1 | 0 | 1 | 2 | 3 |
|---|---|---|---|---|---|---|
| $y_1$ | −8 | −1 | 0 | 1 | 8 | 27 |
| $y_2$ | −18 | −11 | −10 | −9 | −2 | 17 |

   A. $y_2 = x^5$
   B. $y_2 = -x^3$
   C. $y_2 = (-x)^3$
   D. $y_2 = (x-10)^3$
   E. $y_2 = x^3 - 10$

9. **What is the determinant of the matrix shown?** $\begin{bmatrix} 7 & 5 \\ 1 & 4 \end{bmatrix}$
   A. −5
   B. −23
   C. 5
   D. 23
   E. 33

10. **Find the zeros of the function $h(x) = \dfrac{x-9}{x+2}$.**
    A. $\{9\}$
    B. $\{-2\}$
    C. $\left\{-\dfrac{9}{2}\right\}$
    D. $\{-2, 9\}$
    E. This function has no zeros.

11. Find the equation of the graph below.

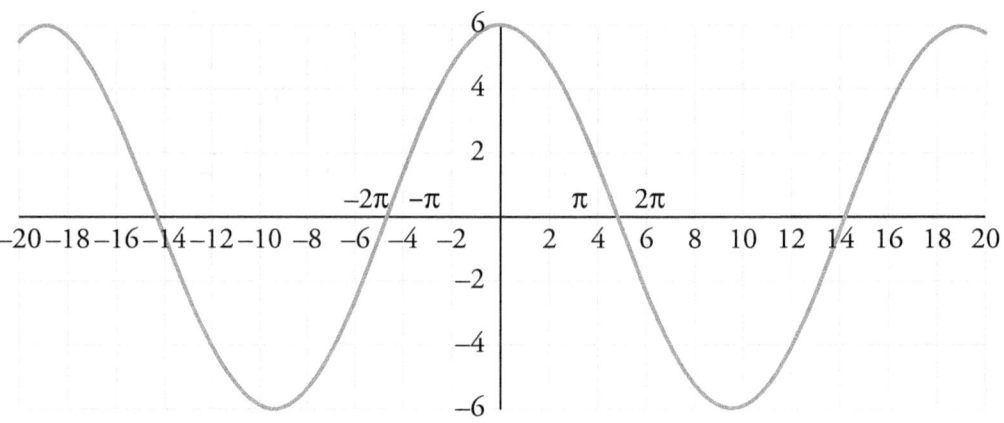

A. $y = 3\cos(3x)$

B. $y = 6\cos(6\pi x)$

C. $y = 6\cos\left(\dfrac{\pi}{3}\right)$

D. $y = 6\sin\left(\dfrac{\pi}{3}\right)$

E. $y = -\sin(3x)$

12. Which of the equations below, when graphed with $y = 10^x$, will show a reflection over the line $y = x$?

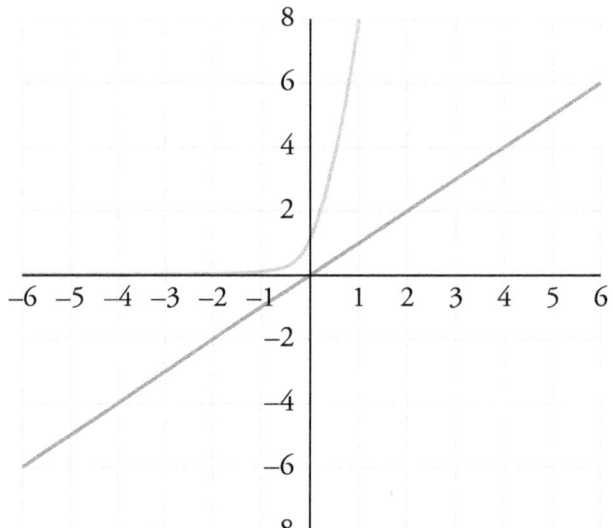

A. $y = (-10)^x$
B. $y = -10^x$
C. $y = 10^{-x}$
D. $y = \log x$
E. $y = \ln x$

13. **Which interval below represents the domain of the function t(x) = sin⁻¹ x?**

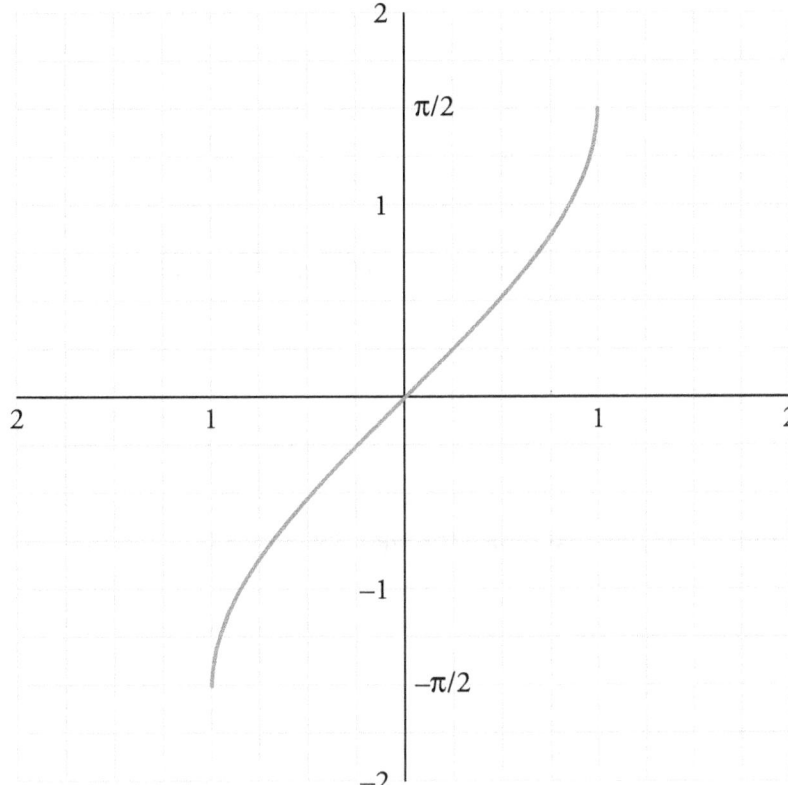

A. $[-1, 1]$
B. $(-1, 1)$
C. $[-\pi, \pi]$
D. $[0, 2\pi]$
E. $[-\infty, \infty]$

14. **Given the function $f(x) = \sqrt{x} - 10$, which translation to the function, shown in the choices below, would ensure the new range values remained greater than zero?**

A. $-f(x)$
B. $f(-x)$
C. $f(x) + 10$
D. $f(x) + 11$
E. None of the above

15. Find the product of the two matrices. $\begin{bmatrix} 2 & 1 \\ 0 & 2 \end{bmatrix} \times \begin{bmatrix} 1 & 1 \\ 2 & 0 \end{bmatrix}$

   A. $\begin{bmatrix} 4 & 2 \\ 4 & 0 \end{bmatrix}$

   B. $\begin{bmatrix} 2 & 1 \\ 0 & 0 \end{bmatrix}$

   C. $\begin{bmatrix} 4 & 0 \\ 0 & 0 \end{bmatrix}$

   D. $\begin{bmatrix} 6 & 3 \\ 4 & 2 \end{bmatrix}$

   E. $\begin{bmatrix} 3 & 2 \\ 2 & 2 \end{bmatrix}$

16. Find the point(s) of intersection of the graphs $f(x) = 4x^2 + 8$ and $g(x) = x^3 + 2x$.
   A. (0, 0) and (0, 8)
   B. (0, 8)
   C. $(2\sqrt{2}, 16)$
   D. (4, 72)
   E. There is no point of intersection.

17. Find the equation of the line that passes through the point (3, 7) and has a slope of $\frac{1}{3}$.

   A. $y = 3x + 7$

   B. $y = \frac{1}{3}x + 7$

   C. $y = \frac{1}{3}x + 6$

   D. $x - 3y + 18 = 0$

   E. Both C and D

18. Find x in the triangle below.

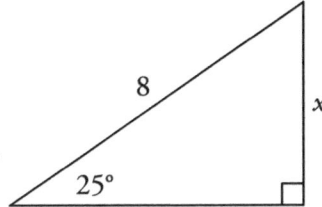

A. 4
B. $4\sqrt{2}$
C. $4\sqrt{3}$
D. 3.381
E. 6.928

19. Find $\left(\cos\dfrac{5\pi}{6}\right)$.

A. $\dfrac{1}{2}$

B. $\dfrac{\sqrt{2}}{2}$

C. $\dfrac{\sqrt{3}}{2}$

D. $-\dfrac{1}{2}$

E. $-\dfrac{\sqrt{3}}{2}$

20. Two friends post a video on the Internet on Monday. On Tuesday each of them forwards the video to three friends. On Wednesday, each of those friends forwards the video to three more friends. The same thing happens Thursday and Friday. How many people have seen the video by the end of Friday?

A. 121
B. 162
C. 240
D. 242
E. 728

21. Which of the following ratios is not equal to $\frac{\sqrt{3}}{3}$?
    A. tan 30°
    B. tan 210°
    C. $\frac{1}{3}$ cot 30°
    D. cot 60°
    E. None of the above (all of the values above equal $\frac{\sqrt{3}}{3}$)

22. The fine for an overdue library book is 50 cents on the first overdue day and increases by 5 cents on each subsequent day. How great a fine is due for a book that has been overdue for 6 days?
    A. $0.30
    B. $0.75
    C. $0.80
    D. $3.00
    E. $3.75

23. Which function below is not continuous over the set of real numbers?

    A. $f(x) = |x|$

    B. $g(x) = \sin x$

    C. $h(x) = \dfrac{1}{x}$

    D. $q(x) = \begin{cases} 0 \text{ for } x > 10 \\ \sqrt{10-x} \text{ for } x \leq 10 \end{cases}$

    E. Both C and D are not continuous.

24. Find a polynomial function with zeros at $-3, -\sqrt{2}, 3, \sqrt{2}$.
    A. $f(x) = 3x^4 - 3x^3 + (\sqrt{2})x^2 - (\sqrt{2})x$
    B. $g(x) = x^2(x-3) + x^3(x-\sqrt{2})$
    C. $h(x) = (x^2 + 9)(x^2 + 2)$
    D. $p(x) = x^4 - 11x^2 + 18$
    E. $t(x) = x^4 - 3x^2 + \sqrt{2}$

25. A certain sound wave can be modeled by the sine function. The wave has an amplitude of 10 and a period of 20. Find a possible equation for the wave.

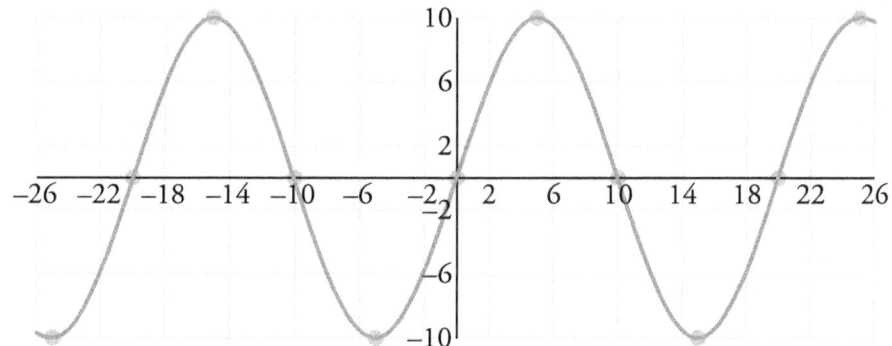

A. $y = \sin 20x + 10$

B. $y = \sin(x - 20) + 10$

C. $y = 10\sin 20x$

D. $y = 10\sin\left(\dfrac{x}{20}\right)$

E. $y = 10\sin\left(\dfrac{\pi x}{10}\right)$

26. Which of the following expressions below is equivalent to $(x - 7)^2$?
   A. $x^2 + 49$
   B. $x^2 - 49$
   C. $x^2 - 14x + 49$
   D. $49x^2$
   E. $2x - 14$

27. **Solve the equation** $\dfrac{|5a-10|}{3} = \dfrac{1}{5}$

   A. $\{47, 53\}$

   B. $\left\{\dfrac{47}{25}, \dfrac{53}{25}\right\}$

   C. $-\dfrac{47}{5}$

   D. $-47$

   E. $53$

28. **Simplify** $\dfrac{x^2+11x+24}{x+8} + \dfrac{1}{x}$

   A. $\dfrac{x^2+3x+1}{x}$

   B. $\dfrac{x^2+11x+4}{2x+3}$

   C. $\dfrac{x^2+11x+9}{3x}$

   D. $2x+4$

   E. $3$

29. **Select the statement below that explains the best first step to solving the equation** $\dfrac{3x-7}{4} = 5x+1.$
   A. Add 7 to both sides.
   B. Subtract $3x$ from both side
   C. Divide both sides by 5.
   D. Multiply both sides by $\dfrac{1}{3}$.
   E. Multiply both sides by 4.

144   SAT Math Level 2

30. Which number line shows the solution to $7x - 5 \geq 9x - 17$?

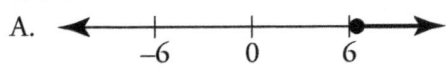

A.
B.
C.
D.
E.

31. A cone and a cylinder have the same volume. The cone has a radius of 14 cm; the cylinder has a radius of 7 cm. The cone has a height of 12 cm. What is the height of the cylinder?
    A. 4 cm
    B. 48 cm
    C. 9 cm
    D. 16 cm
    E. 36 cm

32. Given $g(x) = 2x^2 + 4$, find $g(-3)$.
    A. $-14$
    B. $-8$
    C. 16
    D. 22
    E. 40

33. If $g(x) = x^2 + 9$ and $f(x) = x^2$, find $f(g(x))$.
    A. $x + 3$
    B. $2x^2 + 9$
    C. $x^4 + 9$
    D. $x^4 + 81$
    E. $x^4 + 18x^2 + 81$

34. **Which of the following equations does not represent a function?**
    A. $y = x^2$
    B. $x = y$
    C. $x = y - 5$
    D. $x = 8$
    E. $y = 10$

35. **A canoe leaves a dock and paddles north across a river at 1.5 mph while the river current carries the canoe eastward at 2 mph. How far is the canoe from the dock after 30 minutes?**
    A. 0.75 mi
    B. 1 mi
    C. 1.25 mi
    D. 1.5 mi
    E. 2 mi

36. **Explain the existence of the dotted lines in the graph of g(x) below.**

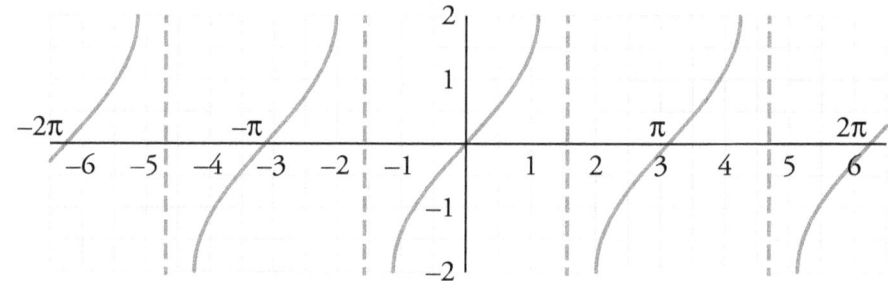

   A. $g(x)$ has dotted lines to show it is periodic.
   B. $g(x)$ is undefined at the dotted lines.
   C. The lines show that $g(x)$ is part linear, part exponential.
   D. The lines show that $g(x)$ has a maximum value occurring at $x = \frac{\pi}{2}$.
   E. The lines prove that $g(x)$ is not a function.

37. In the given graph, if $y_1 = x^2$, then which is the most likely equation to represent $y_2$?

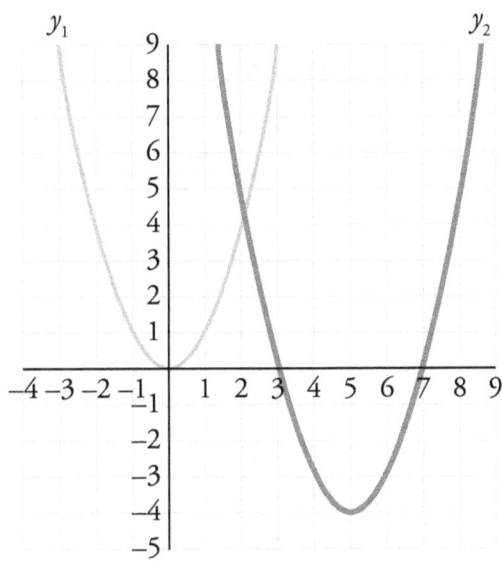

A. $y_2 = 9x^2$
B. $y_2 + 4 = x^2 + 5$
C. $y_2 = (x - 5)^2 - 4$
D. $y_2 = (x - 5)(x - 4)$
E. None of the above

38. Given the piecewise function $f(x) = \begin{cases} x+3 & \text{for } x \geq 0 \\ 5 & \text{for } x < 0 \end{cases}$ find $f(8)$.

A. 2
B. 5
C. 8
D. 11
E. None of the above

39. A kite is flying at a height of 50 feet, 100 feet north and 100 feet east of its owner. If the kite string is taut, how long is the string?

A. 112 feet
B. 141 feet
C. 150 feet
D. 200 feet
E. 250 feet

40. What is the value of x?

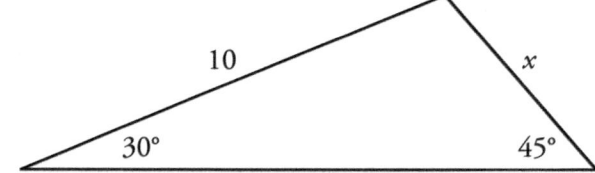

A. 5
B. 6
C. 6.7
D. 7.1
E. 7.5

41. Find the area of the region bounded by $\begin{cases} x^2 + y^2 = 25 \\ x \geq 0 \\ y \geq 0 \end{cases}$

A. 50p

B. 25p

C. $\dfrac{25\pi}{4}$

D. $\dfrac{5\pi}{4}$

E. 25

42. The ages of the participants in a hula-hoop contest are as follows: 10, 18, 22, 17, 77, 19, 13, 20, 10, 15. Which measure would most accurately represent the data as a whole?
A. range
B. interquartile range
C. mode
D. mean
E. median

43. Which equation below represents a parabola with axis of symmetry x = 3?
A. $y^2 = (x - 3)$
B. $y = (x - 3)^2$
C. $x^2 + y^2 = 9$
D. $y = x^2 + 3$
E. $y = 3x^2$

44. If x represents the degree measure of an acute angle of a right triangle, and $\cos x = \frac{15}{17}$ find tax x.

   A. $\frac{17}{15}$

   B. $\frac{8}{15}$

   C. $\frac{2}{17}$

   D. 1

   E. 30°

45. The interquartile range shown in the boxplot is

   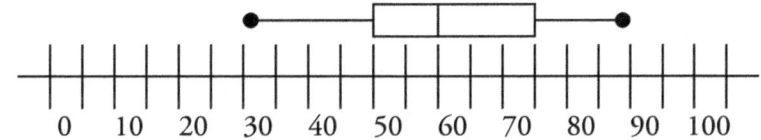

   A. 25
   B. 30
   C. 35
   D. 60
   E. 95

46. Simplify the expression: sin x cos x tan x.
   A. sin x
   B. (sin x)²
   C. sec x
   D. csc x
   E. 1

47. Given the function $f(x) = \begin{cases} x^2 + 5 & \text{for } x > 0 \\ x + 5 & \text{for } x \leq 0 \end{cases}$ for what values of x is the graph increasing?
   A. For $x > 0$
   B. For $x < 0$
   C. For all real values of $x$
   D. For $x > 5$
   E. The graph is only decreasing.

SAT Math Level 2 Practice Test 1

48. Given the following table of values, what are possible equations for y1 and y2?

| $x$ | $y_1$ | $y_2$ |
|---|---|---|
| −3 | −3 | −3 |
| −2 | −2 | −2 |
| −1 | −1 | −1 |
| 0 | 0 | 0 |
| 1 | 1 | −1 |
| 2 | 2 | −2 |
| 3 | 3 | −3 |

A. $y_1 = x, y_2 = -x$
B. $y_1 = x, y_2 = |x|$
C. $y_1 = x, y_2 = -|x|$
D. $y_1 = -x, y_2 = |x|$
E. $y_1 = |x|, y_2 = -|x|$

49. In a card game, you get another turn if the card you draw is red or if it is a jack, queen, king, or ace. You are the first to draw from a full deck of 52 cards. What is the chance you will get another turn?

A. $\dfrac{2}{13}$

B. $\dfrac{4}{13}$

C. $\dfrac{1}{2}$

D. $\dfrac{17}{26}$

E. $\dfrac{21}{26}$

50. Describe the point $(1, \sqrt{3})$ using polar coordinates.
A. $(\sqrt{2}, 30°)$
B. $(\sqrt{3}, 60°)$
C. $(1, 30°)$
D. $(2, 30°)$
E. $(2, 60°)$

## SAT Math 2 Practice Test 1—Answer Key

| | | | | |
|---|---|---|---|---|
| 1. B | 11. C | 21. E | 31. D | 41. C |
| 2. C | 12. D | 22. B | 32. D | 42. E |
| 3. E | 13. A | 23. C | 33. E | 43. B |
| 4. B | 14. D | 24. D | 34. D | 44. B |
| 5. C | 15. A | 25. E | 35. C | 45. A |
| 6. A | 16. D | 26. C | 36. B | 46. B |
| 7. C | 17. E | 27. B | 37. C | 47. C |
| 8. E | 18. D | 28. A | 38. D | 48. C |
| 9. D | 19. E | 29. E | 39. C | 49. D |
| 10. A | 20. D | 30. B | 40. D | 50. E |

# SAT Math Level 2 Practice Test 1

## Answers and Rationales

*Instructions:* For each problem, select the choice that best answers the question.

1. **Simplify (2 - 3i)(4i).**
   A. $-12 + 8i$
   B. $12 + 8i$
   C. $8 - 12i$
   D. $8 + 12i$
   E. $-8 + 12i$

**Answer: B**
Using the Distributive Property, we get $(2 \times 4i) - (3i \times 4i) = 8i - (12i^2) = 8i - (-1)(12) = 12 + 8i$.

2. **Solve $\sqrt{n^2 + 16} = 3n$**
   A. 2
   B. ±2
   C. $\pm\sqrt{2}$
   D. $\pm\dfrac{4}{3}$
   E. No Real Solution

**Answer: C**
Square both sides of the equation.
$$n^2 + 16 = 9n^2$$
$$16 = 8n^2$$
$$2 = n^2$$
Take the plus or minus square root of both sides $\quad \pm\sqrt{2} = n$

3. **Solve for x such that $0 \le x \le 2p$: $\dfrac{1}{2}\sin(2x) - \dfrac{\sqrt{2}}{4} = 0$**
   A. $\left\{\dfrac{\pi}{4}\right\}$
   B. $\left\{\dfrac{\pi}{8}\right\}$
   C. $\left\{\dfrac{\pi}{8}, \dfrac{\pi}{4}\right\}$
   D. $\left\{\pm\dfrac{\pi}{8}, \pm\dfrac{\pi}{4}\right\}$
   E. $\left\{\dfrac{\pi}{8}, \dfrac{3\pi}{8}, \dfrac{9\pi}{8}, \dfrac{11\pi}{8}\right\}$

SAT Math Level 2 Practice Test 1: Answers and Rationales    153

**Answer: E**

Isolate the trigonometric function.

$$\frac{1}{2}\sin(2x) - \frac{\sqrt{2}}{4} = 0$$

$$\frac{1}{2}\sin(2x) = \frac{\sqrt{2}}{4}$$

$$\sin(2x) = \frac{\sqrt{2}}{2}$$

Next list all angles that have a sine value of $\frac{\sqrt{2}}{2}$: $\frac{\pi}{4}, \frac{3\pi}{4}, \frac{9\pi}{4}, \frac{11\pi}{4}, \frac{17\pi}{4}\ldots$

Using the given expression for the angle, $2x$   $2x = \frac{\pi}{4}, \frac{3\pi}{4}, \frac{9\pi}{4}, \frac{11\pi}{4}, \frac{17\pi}{4}\ldots$ then $x = \frac{\pi}{8}, \frac{3\pi}{8}, \frac{9\pi}{8}, \frac{11\pi}{8}, \frac{17\pi}{8}\ldots$

But only the first 4 values listed fall within the given solution specifications, resulting in choice E.

4. **Find the solution to the system of equations.** $\begin{cases} 4x + 2y = 18 \\ y = -2x + 9 \end{cases}$
   A. No solution
   B. Infinitely many solutions
   C. (2, 1)
   D. (9, 18)
   E. (0, 0)

**Answer: B**

These two equations represent the same line, as the first can be algebraically manipulated to match the second:   $4x + 2y = 18$

$2y = -4x + 18$

$y = 2x + 9$

Since they are the same line, their intersection, or solution point, is every point on the infinitely long line, making choice B the correct answer.

5. If $h(x) = \frac{3x+4}{x}$ for all real values of x ≠ 0, find h-1(x).

   A. $h^{-1}(x) = \frac{-3x-4}{x}$

   B. $h^{-1}(x) = \frac{x}{3x+4}$

   C. $h^{-1}(x) = \frac{4}{x-3}$

   D. $h^{-1}(x) = \left(\frac{1}{x}\right)(3x+4)$

   E. None of the above represent $h^{-1}(x)$.

**Answer: C**

Start with $y = \dfrac{3x+4}{x}$ and replace $x$ and $y$ to find the inverse. $\quad x = \dfrac{3y+4}{y}$

Then solve for $y$: $\quad x = \dfrac{3y+4}{y}$

Cross multiply: $\quad xy = 3y + 4$
Put $y$-terms on same side: $\quad xy - 3y = 4$
Factor out a $y$: $\quad y(x-3) = 4$
Divide both sides by $(x-3)$: $\quad y = \dfrac{4}{x-3}$

6. **What values of x will ensure that all values of this function are real numbers?** $f(x) = \sqrt{3-2x}$

   A. $\left\{x \mid x \leq \dfrac{3}{2}\right\}$

   B. $\left\{x \mid x \geq \dfrac{2}{3}\right\}$

   C. $\left\{x \mid x - \dfrac{3}{2} \leq x \leq \dfrac{3}{2}\right\}$

   D. $\left\{x \mid x \neq \dfrac{3}{2}\right\}$

   E. All Real numbers, $x$

**Answer: A**

To keep the function defined over the real numbers, the radicand must not be negative. Algebraically: $3 - 2x \geq 0$. Solving this inequality yields $x \leq \dfrac{3}{2}$, or choice A.

7. **Which of the following would produce an odd number?**
   A. $(Odd + Even + Odd) \times Odd$
   B. $(Odd \times Odd) + Odd$
   C. $(Even + Odd + Even) \times Odd$
   D. $(Odd + Odd) \times (Even + Odd)$
   E. $(Odd \times Odd) - Odd$

**Answer: C**

Answers A and D fail because the quantity inside the opening parentheses is even. Even times odd is even. Answers B and E fail because Odd + Odd or Odd − Odd is an even number. Answer C represents an odd number (in parentheses) multiplied by an odd number.

8. **Based on the given table, if y1 = x3, what is the equation for y2?**

| X | −2 | −1 | 0 | 1 | 2 | 3 |
|---|---|---|---|---|---|---|
| $y_1$ | −8 | −1 | 0 | 1 | 8 | 27 |
| $y_2$ | −18 | −11 | −10 | −9 | −2 | 17 |

A. $y_2 = x^5$
B. $y_2 = -x^3$
C. $y_2 = (-x)^3$
D. $y_2 = (x - 10)^3$
E. $y_2 = x^3 - 10$

**Answer: E**

Each value of $y_2$ is 10 less than the value for $y_1$.

9. **What is the determinant of the matrix shown?** $\begin{bmatrix} 7 & 5 \\ 1 & 4 \end{bmatrix}$
   A. −5
   B. −23
   C. 5
   D. 23
   E. 33

**Answer: D**

The determinant of a matrix $\begin{pmatrix} a & b \\ c & d \end{pmatrix}$ is $ad - bc = 7 \times 4 - 5 \times 1 = 23$.

10. **Find the zeros of the function** $h(x) = \dfrac{x-9}{x+2}$.

   A. {9}
   B. {−2}
   C. $\left\{-\dfrac{9}{2}\right\}$
   D. {−2, 9}
   E. This function has no zeros.

**Answer: A**

When $x = 9$, $h(x) = \dfrac{0}{11} = 0$. −2 is not a zero but a value for which the function is undefined.

11. **Find the equation of the graph below.**

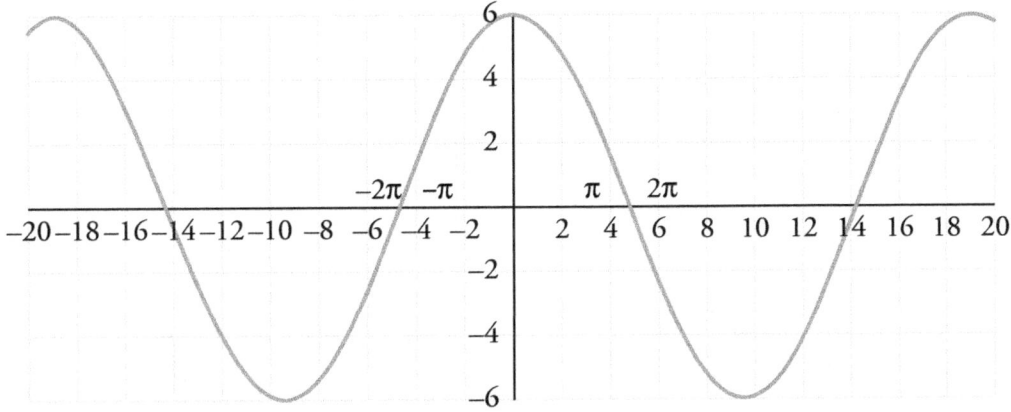

A. $y = 3\cos(3x)$

B. $y = 6\cos(6\pi x)$

C. $y = 6\cos\left(\dfrac{\pi}{3}\right)$

D. $y = 6\sin\left(\dfrac{\pi}{3}\right)$

E. $y = -\sin(3x)$

**Answer: C**

From the diagram, the amplitude of the cosine function has been multiplied by 6, while the period has been lengthened by a factor of 3.

SAT Math Level 2 Practice Test 1: Answers and Rationales

12. Which of the equations below, when graphed with y = 10x, will show a reflection over the line y = x?

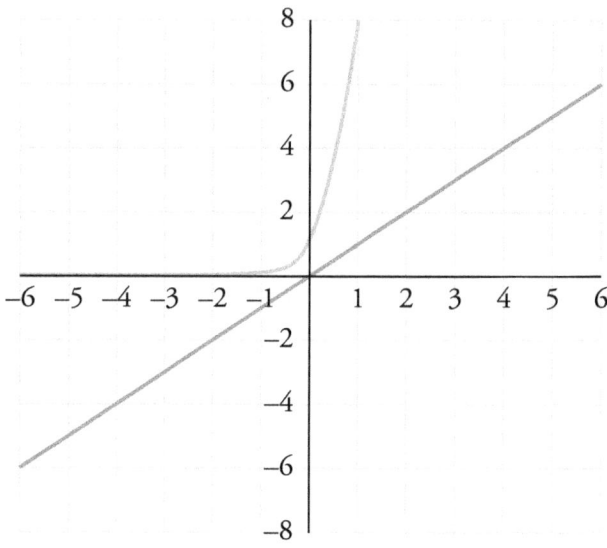

A. $y = (-10)^x$
B. $y = -10^x$
C. $y = 10^{-x}$
D. $y = \log x$
E. $y = \ln x$

**Answer: D**

The function that will be reflected over $x = y$ is the inverse function. $y = \log x$ is the inverse function of $y = 10^x$.

13. **Which interval below represents the domain of the function t(x) = sin-1 x?**

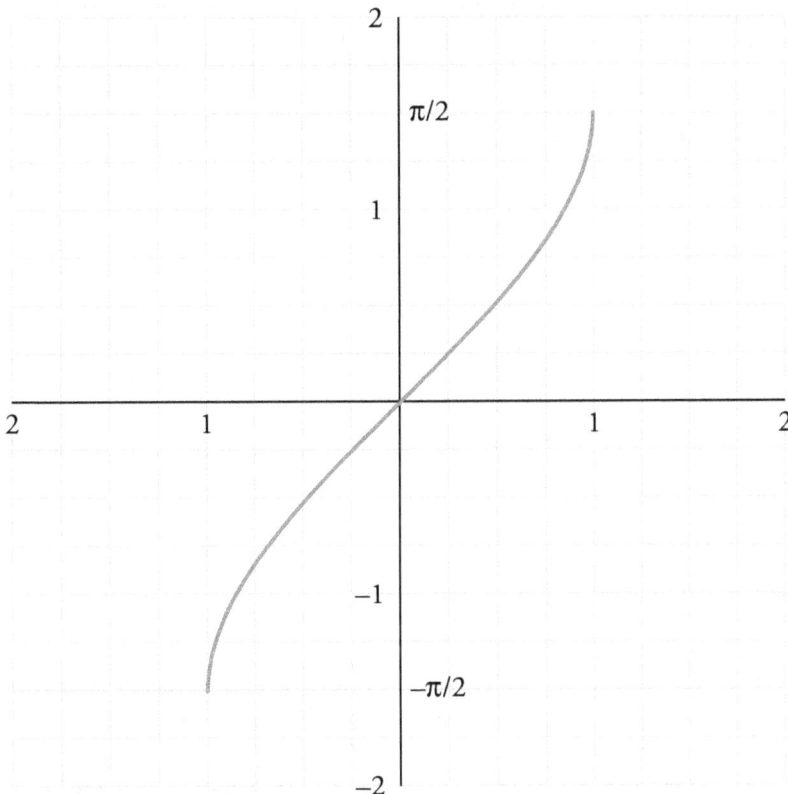

A. [−1, 1]
B. (−1, 1)
C. [−π, π]
D. [0, 2π]
E. [−∞, ∞]

**Answer: A**
Each input is a value of the sine function, whose range is between −1 and 1, inclusive.

14. **Given the function $f(x) = \sqrt{x} - 10$, which translation to the function, shown in the choices below, would ensure the new range values remained greater than zero?**
    A. −f(x)
    B. f(−x)
    C. f(x) + 10
    D. f(x) + 11
    E. None of the above

**Answer: D**
The location of the minimum of $f(x) = \sqrt{x} - 10$ is (0, −10). Choice C would raise this point only up to (0,0). Therefore choice D, which raises the minimum to (0,1), is the translation that would keep range values positive. Be advised that A is not a valid choice, since the x-intercept of the original f(x) graph is

SAT Math Level 2 Practice Test 1: Answers and Rationales  159

(100,0). This point, while far outside the standard viewing range, represents the point where $f(x)$ switches from negative values to positive. Choice A still contains this point. In other words, $-f(x)$ is a graph that does not have a range completely greater than zero.

15. **Find the product of the two matrices.** $\begin{bmatrix} 2 & 1 \\ 0 & 2 \end{bmatrix} \times \begin{bmatrix} 1 & 1 \\ 2 & 0 \end{bmatrix}$

   A. $\begin{bmatrix} 4 & 2 \\ 4 & 0 \end{bmatrix}$

   B. $\begin{bmatrix} 2 & 1 \\ 0 & 0 \end{bmatrix}$

   C. $\begin{bmatrix} 4 & 0 \\ 0 & 0 \end{bmatrix}$

   D. $\begin{bmatrix} 6 & 3 \\ 4 & 2 \end{bmatrix}$

   E. $\begin{bmatrix} 3 & 2 \\ 2 & 2 \end{bmatrix}$

Answer: A

Multiply the rows of the first matrix by the columns of the second matrix, in each case adding the two products: $2 \times 1 + 1 \times 2 = 4$, $2 \times 1 + 1 \times 0 = 2$, $0 \times 1 + 2 \times 2 = 4$, $0 \times 1 + 2 \times 0 = 0$

16. **Find the point(s) of intersection of the graphs $f(x) = 4x^2 + 8$ and $g(x) = x^3 + 2x$.**

    A. $(0, 0)$ and $(0, 8)$
    B. $(0, 8)$
    C. $(2\sqrt{2}, 16)$
    D. $(4, 72)$
    E. There is no point of intersection.

Answer: D

Algebraically, find the x-coordinate of intersection by setting the two function expressions equal to each other. $\qquad 4x^2 + 8 = x^3 + 2x$
Put all terms on same side equal to zero. $\qquad 0 = x^3 + 2x - 4x^2 - 8$
Factor by grouping. $\qquad 0 = x(x^2 + 2) - 4(x^2 + 2)$
$\qquad 0 = (x - 4)(x^2 + 2)$
Set each factor equal to zero. $\qquad x - 4 = 0$ or $x^2 + 2 = 0$
This scenario yields only one real solution for $x$: $x = 4$. Evaluate either function to find $y$: $y = 4(4)^2 + 8 = 72$. Alternatively, use graphing technology to graph the two equations and find their intersection, realizing that the intersection point exists despite the fact that it may appear outside the standard viewing window.

17. **Find the equation of the line that passes through the point (3, 7) and has a slope of $\frac{1}{3}$.**

    A. $y = 3x + 7$

160  SAT Math Level 2

B. $y = \frac{1}{3}x + 7$

C. $y = \frac{1}{3}x + 6$

D. $x - 3y + 18 = 0$

E. Both C and D

**Answer: E**

First find the equation of the line using point-slope form: $y - y_1 = m(x - x_1)$.
$$y - 7 = \frac{1}{3}(x - 3)$$
$$y - 7 = \frac{1}{3}x - 1$$
$$y = \frac{1}{3}x + 6$$

But this equation can be rewritten, first by multiplying through by 3:
$$3y = x + 18$$
$$0 = x - 3y + 18$$
$$x - 3y + 18 = 0$$

So both C and D are correct.

18. **Find x in the triangle below.**

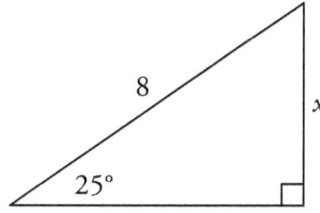

A. 4
B. $4\sqrt{2}$
C. $4\sqrt{3}$
D. 3.381
E. 6.928

**Answer: D**

The right triangle comes with information representing the sine ratio:

$\sin\theta = \dfrac{opposite}{hypotenuse}$, Using the given information,

$\sin 25° = \dfrac{x}{8}$
$x = 8\sin 25°$
$x \approx 3.381$

19. **Find** $\left(\cos\dfrac{5\pi}{6}\right)$.

A. $\dfrac{1}{2}$
B. $\dfrac{\sqrt{2}}{2}$
C. $\dfrac{\sqrt{3}}{2}$
D. $-\dfrac{1}{2}$
E. $-\dfrac{\sqrt{3}}{2}$

**Answer: E**

In terms of the unit circle, the angle in question is in the second quadrant, where the cosine is negative. The reference angle at that location is $\dfrac{\pi}{6}$ or 30° and $\cos 30 = \dfrac{\sqrt{3}}{2}$. Combining these two conditions results in choice E.

20. **Two friends post a video on the Internet on Monday. On Tuesday each of them forwards the video to three friends. On Wednesday, each of those friends forwards the video to three more friends. The same thing happens Thursday and Friday. How many people have seen the video by the end of Friday?**

A. 121
B. 162
C. 240
D. 242
E. 728

**Answer: D**

The total number of viewers is $2 + (2 \times 3) + (2 \times 3^2) + (2 \times 3^3) + (2 \times 3^4) = 242$.

21. **Which of the following ratios is not equal to $\frac{\sqrt{3}}{3}$?**

    A. $\tan 30°$
    
    B. $\tan 210°$
    
    C. $\frac{1}{3} \cot 30°$
    
    D. $\cot 60°$
    
    E. None of the above (all of the values above equal $\frac{\sqrt{3}}{3}$)

**Answer: E**

Unit circle relationships find all the choices to be equal.

For example: $\sin 30° = \frac{1}{2}, \cos 30° = \frac{\sqrt{3}}{2}$

$$\tan 30° = \frac{\sin 30°}{\cos 30°} = \frac{\frac{1}{2}}{\frac{\sqrt{3}}{2}} = \frac{1}{\sqrt{3}} = \frac{\sqrt{3}}{3}, \quad \left(\frac{1}{3}\right)\cot 30° = \left(\frac{1}{3}\right)\frac{\cos 30°}{\sin 30°} = \left(\frac{1}{3}\right)\frac{\frac{\sqrt{3}}{2}}{\frac{1}{2}} = \frac{\sqrt{3}}{3}$$

22. **The fine for an overdue library book is 50 cents on the first overdue day and increases by 5 cents on each subsequent day. How great a fine is due for a book that has been overdue for 6 days?**

    A. $0.30
    B. $0.75
    C. $0.80
    D. $3.00
    E. $3.75

**Answer: B**

The total fine is $.50 for the first day and $5 \times \$0.05$ for the remaining days, adding to $0.75.

23. **Which function below is not continuous over the set of real numbers?**

    A. $f(x) = |x|$
    
    B. $g(x) = \sin x$

SAT Math Level 2 Practice Test 1: Answers and Rationales

C. $h(x) = \dfrac{1}{x}$

D. $q(x) = \begin{cases} 0 \text{ for } x > 10 \\ \sqrt{10-x} \text{ for } x \leq 10 \end{cases}$

E. Both C and D are not continuous.

Answer: C

The graph of $h(x) = \dfrac{1}{x}$ is undefined at $x = 0$, so the function is not continuous. While the equation listed in choice D is a piecewise function, the fact that the values at the "split" point are equal keeps the function continuous.

24. **Find a polynomial function with zeros at -3, -√2, 3, √2.**
    A. $f(x) = 3x^4 - 3x^3 + (\sqrt{2})x^2 - (\sqrt{2})x$
    B. $g(x) = x^2(x-3) + x^3(x-\sqrt{2})$
    C. $h(x) = (x^2+9)(x^2+2)$
    D. $p(x) = x^4 - 11x^2 + 18$
    E. $t(x) = x^4 - 3x^2 + \sqrt{2}$

Answer: D

The zero of a function is the same as the root of an equation. If r is a root of a polynomial equation then $(x - r)$ is a factor. Use the 4 given zeros, or roots, to create 4 factors:

$$(x+3)(x-3)(x+\sqrt{2})(x-\sqrt{2})$$

After multiplying the conjugate pairs: $(x^2 - 9)(x^2 - 2)$

After multiplying the binomials: $x^4 - 9x^2 - 2x^2 + 18$, which simplifies to choice D.

25. A certain sound wave can be modeled by the sine function. The wave has an amplitude of 10 and a period of 20. Find a possible equation for the wave.

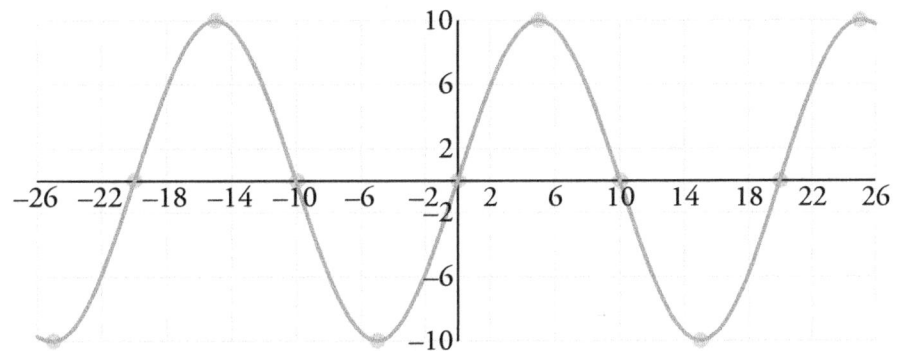

A. $y = \sin 20x + 10$

B. $y = \sin(x - 20) + 10$

C. $y = 10\sin 20x$

D. $y = 10\sin\left(\dfrac{x}{20}\right)$

E. $y = 10\sin\left(\dfrac{\pi x}{10}\right)$

**Answer: E**

Start with the basic formula $y = A\sin(nx)$ where A = the amplitude and n is the period factor. To find $n$, use the relationship: Standard period = $n$(Changed period).

$$2\pi = n(20)$$
$$n = \dfrac{2\pi}{20} = \dfrac{\pi}{10}$$

26. Which of the following expressions below is equivalent to $(x - 7)^2$?

   A. $x^2 + 49$
   B. $x^2 - 49$
   C. $x^2 - 14x + 49$
   D. $49x^2$
   E. $2x - 14$

SAT Math Level 2 Practice Test 1: Answers and Rationales

**Answer: C**

The squared binomial can be expanded: $(x-7)(x-7)$
And multiplied: $x^2 - 7x - 7x + 49$
And combined: $x^2 - 14x + 49$

**27. Solve the equation** $\dfrac{|5a-10|}{3} = \dfrac{1}{5}$

A. $\{47, 53\}$

B. $\left\{\dfrac{47}{25}, \dfrac{53}{25}\right\}$

C. $-\dfrac{47}{5}$

D. $-47$

E. $53$

**Answer: B**

First, isolate the absolute value bars by multiplying both sides of the equation by 3.

$$|5a-10| = \dfrac{3}{5}$$

Then set up two equations to represent the definition of absolute value.

$$5a - 10 = \dfrac{3}{5} \quad \text{and} \quad 5a - 10 = -\dfrac{3}{5}$$
$$5a = \dfrac{53}{5} \quad \text{and} \quad 5a = \dfrac{47}{5}$$
$$a = \dfrac{53}{25} \quad \text{and} \quad a = \dfrac{47}{25}$$

**28. Simplify** $\dfrac{x^2 + 11x + 24}{x+8} + \dfrac{1}{x}$

A. $\dfrac{x^2 + 3x + 1}{x}$

B. $\dfrac{x^2 + 11x + 4}{2x + 3}$

C. $\dfrac{x^2 + 11x + 9}{3x}$

D. $2x + 4$

E. $3$

166    SAT Math Level 2

**Answer: A**

First factor and cancel in the first portion of the expression.

$$\frac{x^2+11x+24}{x+8} \to \frac{(x+8)(x+3)}{(x+8)} \to x+3$$

Then find common denominators to add the two expressions together.

$$(x+3)+\frac{1}{x}$$
$$\frac{x(x+3)}{x}+\frac{1}{x}$$
$$\frac{x^2+3x+1}{x}$$

29. **Select the statement below that explains the best first step to solving the equation $\frac{3x-7}{4}=5x+1$.**
    A. Add 7 to both sides.
    B. Subtract $3x$ from both side
    C. Divide both sides by 5.
    D. Multiply both sides by $\frac{1}{3}$.
    E. Multiply both sides by 4.

**Answer: E**

While choices A–D do describe steps that may take place later in the equation-solving process, they are not advisable as first steps due to the interference of the denominator on the left side of the equation. Choice E as a first step "clears" that denominator so the rest of the steps can occur.

30. **Which number line shows the solution to $7x - 5 \geq 9x - 17$?**

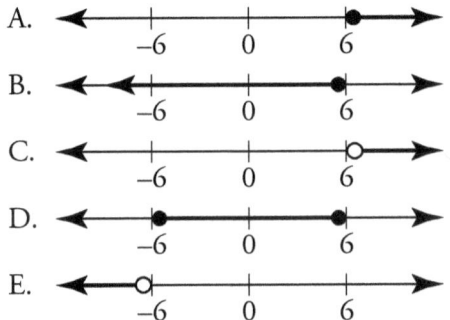

**Answer: B**

First gather all the $x$-terms on one side of the inequality and the numbers on the other.
$$7x - 5 \geq 9x - 17$$
$$-2x \geq -12$$

Dividing both sides of an inequality by a negative number reverses the inequality sign. So division by −2 on both sides results in $x \leq 6$, which is graphed in choice B.

SAT Math Level 2 Practice Test 1: Answers and Rationales

31. **A cone and a cylinder have the same volume. The cone has a radius of 14 cm; the cylinder has a radius of 7 cm. The cone has a height of 12 cm. What is the height of the cylinder?**
   A. 4 cm
   B. 48 cm
   C. 9 cm
   D. 16 cm
   E. 36 cm

Answer: D

The volume of a cone equals the area of the base times one-third the height:
$$V = b \times \frac{h}{3}$$
$$= \pi r^2 \left(\frac{h}{3}\right)$$
$$= 14^2 \pi \left(\frac{12}{3}\right)$$
$$= 784\pi \text{ cm}^3$$

The volume of a cylinder equals the area of the base times the height:
$$V = b \times h$$
$$= \pi r^2 h$$
$$= 7^2 \pi h$$
$$= 49 h \pi \text{ cm}^3$$

If the volumes are equal, then $49\pi h = 784\pi$. Therefore, $h = 16$.

32. **Given $g(x) = 2x^2 + 4$, find $g(-3)$.**
   A. −14
   B. −8
   C. 16
   D. 22
   E. 40

Answer: D

$g(-3) = 2(-3)^2 + 4 = 2(9) + 4 = 22$  G(−3) = 2(−3)^2 + 4 = 2(9) + 4 = 22

33. **If $g(x) = x^2 + 9$ and $f(x) = x^2$, find $f(g(x))$.**
   A. $x + 3$
   B. $2x^2 + 9$
   C. $x^4 + 9$
   D. $x^4 + 81$
   E. $x^4 + 18x^2 + 81$

Answer: E

Evaluate: $f(g(x)) = f(x^2 + 9) = (x^2 + 9)^2$

Simplify by expanding:

$$(x^2 + 9)(x^2 + 9)$$
$$x^4 + 9x^2 + 9x^2 + 81$$
$$x^4 + 18x^2 + 81$$

**34. Which of the following equations does not represent a function?**
A. $y = x^2$
B. $x = y$
C. $x = y - 5$
D. $x = 8$
E. $y = 10$

Answer: D

A function can have only one output, $y$, for each input, $x$. A table of ordered pairs for choice D could be

| $x$ | 8 | 8 | 8 | 8 | 8 |
|---|---|---|---|---|---|
| $y$ | −2 | 0 | 1 | 3 | 6 |

This shows that the only $x$-input, 8, has multiple $y$-outputs. Therefore, the equation is not a function.

**35. A canoe leaves a dock and paddles north across a river at 1.5 mph while the river current carries the canoe eastward at 2 mph. How far is the canoe from the dock after 30 minutes?**
A. 0.75 mi
B. 1 mi
C. 1.25 mi
D. 1.5 mi
E. 2 mi

Answer: C

The motion of the canoe can be described by vectors: a northward vector of 1.5 mph created by the paddling and an eastward vector of 2 mph created by the river current.
Use the Pythagorean Theorem to find the length of the vector representing the sum of the two vectors: $1.5^2 + 2^2 = c^2$. Then c is a vector moving roughly northeast at 2.5 mph.

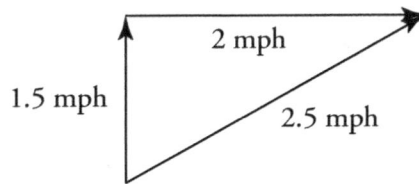

After moving away from the dock at a rate of 2.5 mph for half an hour, the canoe will be 1.25 mi distant from the dock.

36. **Explain the existence of the dotted lines in the graph of g(x) below.**

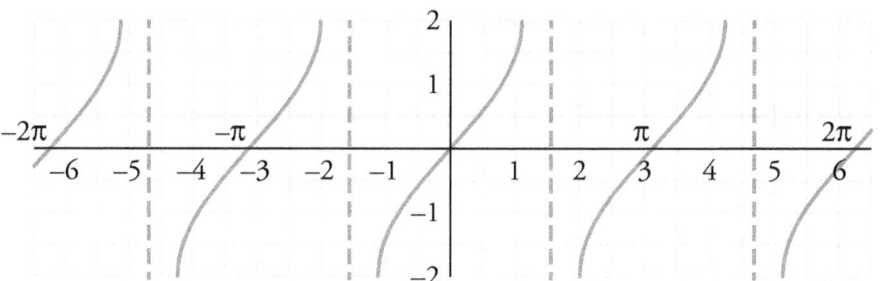

A. g(x) has dotted lines to show it is periodic.
B. g(x) is undefined at the dotted lines.
C. The lines show that g(x) is part linear, part exponential.
D. The lines show that g(x) has a maximum value occurring at $x = \frac{\pi}{2}$.
E. The lines prove that g(x) is not a function.

**Answer: B**

Dotted lines in a graph represent asymptotes, which indicate places where a graph is trending, but will never reach. In this case, g(x) appears to be the tangent function. $\tan\left(\frac{\pi}{2}\right)$ is undefined. The curve will stretch infinitely high as x approaches $\frac{\pi}{2}$ but will have no value for that input of the function.

37. **In the given graph, if y1 = x2, then which is the most likely equation to represent y2?**

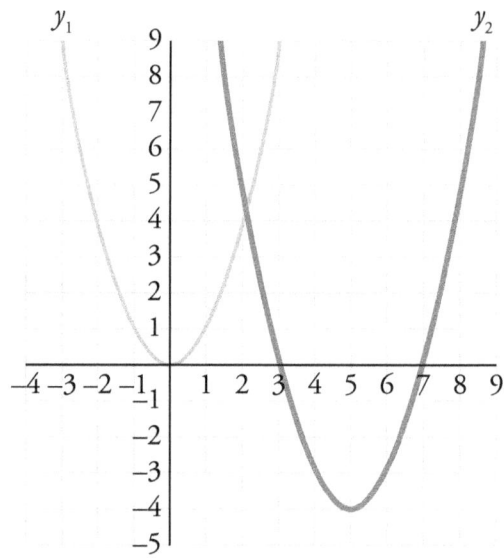

170    SAT Math Level 2

A. $y_2 = 9x^2$
B. $y_2 + 4 = x^2 + 5$
C. $y_2 = (x - 5)^2 - 4$
D. $y_2 = (x - 5)(x - 4)$
E. None of the above

**Answer: C**

The vertex of the graph of $y_2$ is (5, −4). Therefore choice C is the best equation.

38. **Given the piecewise function** $f(x) = \begin{cases} x+3 & \text{for } x \geq 0 \\ 5 & \text{for } x < 0 \end{cases}$ **find f(8).**
    A. 2
    B. 5
    C. 8
    D. 11
    E. None of the above

**Answer: D**

The input value, 8, is greater than zero, so the function is to be evaluated using the first portion of the rule: $f(x) = x + 3$, so $f(8) = 8 + 3 = 11$.

39. **A kite is flying at a height of 50 feet, 100 feet north and 100 feet east of its owner. If the kite string is taut, how long is the string?**
    A. 112 feet
    B. 141 feet
    C. 150 feet
    D. 200 feet
    E. 250 feet

**Answer: C**

The height, the distance north, and the distance east can be seen as forming a cube. The kite string forms a 3-dimensional diagonal from the bottom left vertex in front to the upper right vertex in the rear.

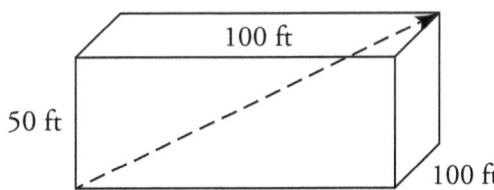

By the Pythagorean theorem, the length of the string is $\sqrt{50^2 + 100^2 + 100^2} = \sqrt{22,500} = 150$ ft.

**40.** What is the value of x?

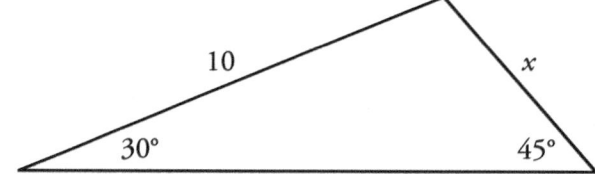

A. 5
B. 6
C. 6.7
D. 7.1
E. 7.5

Answer: D

By the law of sines, the sides of a triangle are proportional to the sines of the opposite angles.

$$\frac{10}{\sin 45°} = \frac{x}{\sin 30°}$$

$$\frac{10(\sin 30°)}{\sin 45°} = x$$

$$x = 10\left(\frac{1}{2}\right) \div \frac{\sqrt{2}}{2}$$

$$= 5 \times \frac{2}{\sqrt{2}}$$

$$= 5\sqrt{2}$$

$$\approx 7.1$$

**41.** Find the area of the region bounded by $\begin{cases} x^2 + y^2 = 25 \\ x \geq 0 \\ y \geq 0 \end{cases}$

A. 50p

B. 25p

C. $\frac{25\pi}{4}$

D. $\frac{5\pi}{4}$

E. 25

Answer: C

The given region is the first quadrant portion of a circle, centered on the origin, with radius 5.

The full area of such a circle would be $A = \pi r^2 = 25\pi$. The first quadrant section is a quarter of the entire circle, resulting in answer choice C.

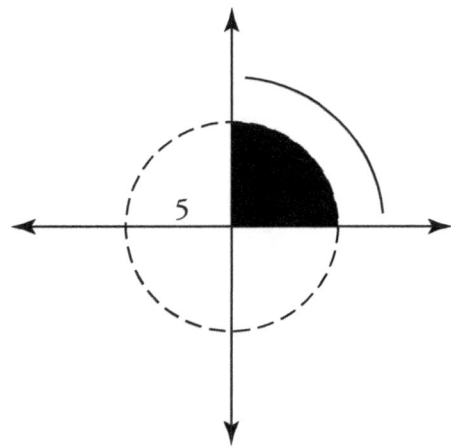

42. **The ages of the participants in a hula-hoop contest are as follows: 10, 18, 22, 17, 77, 19, 13, 20, 10, 15. Which measure would most accurately represent the data as a whole?**
    A. range
    B. interquartile range
    C. mode
    D. mean
    E. median

Answer: E

The data set contains an outlier. The range including the outlier would represent too wide a variation of the data, most of which falls within a narrow range. The interquartile range would show that the data was mostly within a narrow range, but would indicate nothing about the general size of the data values. The mode would be a value smaller than all the other data items. The mean would be too large, skewed by the presence of the outlier. Only the median would have a value representative of the data as a whole.

43. **Which equation below represents a parabola with axis of symmetry x = 3?**
    A. $y^2 = (x - 3)$
    B. $y = (x - 3)^2$
    C. $x^2 + y^2 = 9$
    D. $y = x^2 + 3$
    E. $y = 3x^2$

Answer: B

Choice B represents a parabola, opening up, with vertex (3, 0). Such a parabola has a line of symmetry at $x = 3$.

44. If x represents the degree measure of an acute angle of a right triangle, and $\cos x = \frac{15}{17}$ find tax x.

   A. $\frac{17}{15}$

   B. $\frac{8}{15}$

   C. $\frac{2}{17}$

   D. 1

   E. 30°

Answer: B

The ratio can be found without ever knowing the value of angle x. First, draw a right triangle with a hypotenuse of 17. Arbitrarily choose one of the angles to be x, then make the leg adjacent to that side have a length of 15. According to the Pythagorean Theorem, the opposite leg must then be 8. Therefore the tangent ratio, "opposite over adjacent," is $\frac{8}{15}$.

45. The interquartile range shown in the boxplot is

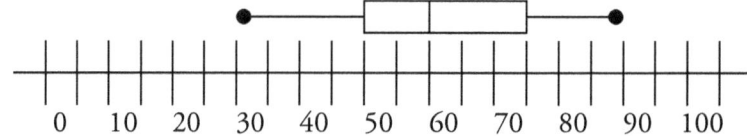

   A. 25
   B. 30
   C. 35
   D. 60
   E. 95

Answer: A

In a box plot, the interquartile range runs from the left side of the box to the right side of the box, in this case from 55 to 80.

46. **Simplify the expression: sin x cos x tan x.**
   A. sin $x$
   B. $(\sin x)^2$
   C. sec $x$
   D. csc $x$
   E. 1

Answer: B

$$\frac{\sin x}{1} \times \frac{\cos x}{1} \times \frac{\sin x}{\cos x} = (\sin x)^2$$

47. **Given the function** $f(x) = \begin{cases} x^2 + 5 & \text{for } x > 0 \\ x + 5 & \text{for } x \leq 0 \end{cases}$ **for what values of x is the graph increasing?**
   A. For $x > 0$
   B. For $x < 0$
   C. For all real values of $x$
   D. For $x > 5$
   E. The graph is only decreasing.

Answer: C

When looking at the graph from left to right, it is first a line with a positive slope, making an increasing graph. The graph is continuous at $x = 0$, since the point (0, 5) is common to both equations. Then, as the graph continues to the right of the y-axis, it is the right half of a parabola, which is also an increasing graph.

48. **Given the following table of values, what are possible equations for $y_1$ and $y_2$?**

| $x$ | $y_1$ | $y_2$ |
|---|---|---|
| −3 | −3 | −3 |
| −2 | −2 | −2 |
| −1 | −1 | −1 |
| 0 | 0 | 0 |
| 1 | 1 | −1 |
| 2 | 2 | −2 |
| 3 | 3 | −3 |

   A. $y_1 = x, y_2 = -x$
   B. $y_1 = x, y_2 = |x|$
   C. $y_1 = x, y_2 = -|x|$
   D. $y_1 = -x, y_2 = |x|$
   E. $y_1 = |x|, y_2 = -|x|$

**Answer: C**

The equation for $y_1$ keeps all the $x$-inputs the same, while the $y_2$ makes all the outputs negative. Choice C accomplishes these transformations.

49. **In a card game, you get another turn if the card you draw is red or if it is a jack, queen, king, or ace. You are the first to draw from a full deck of 52 cards. What is the chance you will get another turn?**

    A. $\dfrac{2}{13}$

    B. $\dfrac{4}{13}$

    C. $\dfrac{1}{2}$

    D. $\dfrac{17}{26}$

    E. $\dfrac{21}{26}$

**Answer: D**

The cards that would give you another turn include all 26 red cards, plus the two black jacks, queens, kings, and aces, making a total of 34 out of 52 or $\dfrac{17}{26}$.

50. **Describe the point $(1, \sqrt{3})$ using polar coordinates.**

    A. $(\sqrt{2}, 30°)$
    B. $(\sqrt{3}, 60°)$
    C. $(1, 30°)$
    D. $(2, 30°)$
    E. $(2, 60°)$

**Answer: E**

A line from the origin to the point $(1, \sqrt{3})$ has a length of 2 and forms a 60° angle with the $x$-axis.

# SECTION IV:
# SAT Math Level 2
# Practice Test 2

# SAT Math Level 2 Practice Test 2

*Sample Test Questions*

*Multiple Choice:* select the best answer to solve the problem.

1. Which of the following expressions is not equivalent to $\frac{a+b}{c}$?

    A. $\frac{1}{c}(a+b)$

    B. $\frac{a}{c}+\frac{b}{c}$

    C. $\frac{b+a}{c}$

    D. $a+b \div c$

    E. All of the above are equivalent to the given expression.

2. The dosage of a certain antibiotic must be measured as 40 mg of medicine for every 25 lb of patient weight. How many milligrams must be prescribed for a 140-lb patient?
    A. 5 mg
    B. 165 mg
    C. 180 mg
    D. 224 mg
    E. 250 mg

3. How many different committees can be formed by selecting 4 members from a pool of 50 candidates?
    A. 5,527,200
    B. 230,300
    C. 200
    D. 54
    E. 4

4. Find $\sum_{n=1}^{5} n^2$
    A. 24
    B. 25
    C. 55
    D. 100
    E. None of the above

5. If n represents any whole number, which expression below represents the product of 2 consecutive, odd numbers?
   A. $(2n + 1)(2n+3)$
   B. $(n + 1)(n + 3)$
   C. $(3n)(5n)$
   D. $n(n^2)$
   E. $n(n+2)$

6. Simplify $2^3 i^5$
   A. $8i$
   B. $-8$
   C. $6i$
   D. $-6i$
   E. $\sqrt{-8}$

7. Find the magnitude of the vector 6i + 8j.
   A. $8\sqrt{3}$
   B. 10
   C. 14
   D. 48
   E. None of the above

8. Write an expression representing the following relationship: "double the square of a number."
   A. $2n^2$
   B. $(2n)^2$
   C. $2(25n)$
   D. $n^4$
   E. Both A and B

9. Simplify the expression $(4x^8 y^5)(4xy^3)^{-2}$
   A. $x^7 y^2$
   B. $-16x^6 y$
   C. $\dfrac{x^4}{12y}$
   D. $\dfrac{x^6}{4y}$
   E. $\dfrac{x^6 y}{4}$

180   SAT Math Level 2

10. **Solve for x:** $\frac{1}{3}x + 2 = \frac{3}{5}x + \frac{1}{3}$

    A. $\frac{2}{3}$

    B. $\frac{4}{5}$

    C. $\frac{25}{4}$

    D. $\frac{35}{3}$

    E. $\frac{48}{5}$

11. **Solve over the complex numbers:** $9x^2 + 49 = 0$

    A. $\pm 2\sqrt{10}$

    B. $\pm \frac{7i}{3}$

    C. $\frac{7}{3}$

    D. $\frac{49}{9}$

    E. $-40$

12. **Which system of equations below has an infinite number of solutions?**

    A. $\begin{cases} 5x + y = 8 \\ 3x - 4y = 14 \end{cases}$

    B. $\begin{cases} 2x + y = 7 \\ y = 4 \end{cases}$

    C. $\begin{cases} 3x - 2y = 8 \\ 6x - 4y = 8 \end{cases}$

    D. $\begin{cases} x + y = 12 \\ 5x + 5y = 60 \end{cases}$

    E. $\begin{cases} x^2 + y^2 = 25 \\ x + y = 5 \end{cases}$

13. **Solve for x: x3 - 3x2 - 3x + 9 = 0**
    A. 0, 3
    B. −2, 2, 3
    C. −√3, √3, 3
    D. −3, 3, 9
    E. 0, 3, 9

14. **Solve for x. Round the answer to the nearest hundredth. 3x = 12**
    A. 2.26
    B. 3.14
    C. 4.00
    D. 4.12
    E. 6.00

15. **Solve the compound inequality: 7 ≤ 3x + 1 ≤ 49**
    A. $0 \le x \le 2$
    B. $2 \le x \le 16$
    C. $3.5 \le x \le 24.5$
    D. $x \le 2$ or $x \le 16$
    E. $x \ge 0$ or $x \le 2$

16. **Solve for x, such that 0° ≤ x ≤ 360°: 8 sin x + 1 = 5**
    A. $\frac{1}{2}$
    B. 1.73
    C. 30
    D. 150
    E. Both C and D

17. **Which function listed below is not defined for x = 0, -2?**
    A. $f(x) = x^2 + 2x$
    B. $g(x) = \sqrt{x^2 + 2x}$
    C. $p(x) = x(x - 2)$
    D. $q(x) = 5$
    E. $h(x) = \dfrac{3}{x^2 + 2x}$

18. If $h(x) = x^2$ and $g(x) = x + 3$, which statement below is false?

    A. $g(h(x)) = x^2 + 3$

    B. $h(g(x)) = x^2 + 9$

    C. $(h \circ g)(x) = x^2 + 6x + 9$

    D. $h(x) \cdot g(x) = x^3 + 3x^2$

    E. B and D are both false

19. Which equation below represents a function with zeros −1, 2, and 7?

    A. $f(x) = x^7 + x^2 - x$
    B. $g(x) = x^2 + 7x - 1$
    C. $h(x) = 2x + 7$
    D. $m(x) = 2x^3 - x^2 + 7x$
    E. $p(x) = x^3 - 8x^2 + 5x + 14$

20. Find the Cartesian equation that corresponds to the given set of parametric equations.

    $\begin{cases} x(t) = 5t \\ y(t) = t^2 + 1 \end{cases}$

    A. $y = x^2 + 5x + 1$

    B. $y = 5x^2 + 5$

    C. $y = \dfrac{x^2}{25} + 1$

    D. $y = 25x^2 + 1$

    E. $y = \dfrac{1}{5}x$

21. If $f(x) = \begin{cases} 2x \text{ for } x < 0 \\ \sqrt{x} \text{ for } x \geq 0 \end{cases}$, find f(-16)

    A. −32
    B. −16
    C. 0
    D. 4
    E. $4i$

22. **If h(x) and g(x) are inverses of each other, then which statement below is true?**
    A. $h(x) = -g(x)$
    
    B. $h(x) = \dfrac{1}{g(x)}$
    
    C. $h(g(a)) = a$
    
    D. $g(h(a)) = a$
    
    E. Both C and D are true

23. **Simplify** $\left(64x^{10}y^{-2}\right)^{3/2}$
    
    A. $\sqrt{64x^{10}y^2}$
    
    B. $8x^5y\sqrt{3}$
    
    C. $96x^{15}y^3$
    
    D. $-512x^{15}y^3$
    
    E. $\dfrac{512x^{15}}{y^3}$

24. **Solve for x:** $x^2 + 4x + 5 = 0$
    
    A. $\{4, 5\}$
    
    B. $\{1, 5\}$
    
    C. $\{2, \sqrt{5}\}$
    
    D. $\dfrac{4 \pm \sqrt{26}}{2}$
    
    E. No real solution

25. **Solve** $x - 5 = \sqrt{x+7}$
    A. $\{5, 7\}$
    B. $\{2, 9\}$
    C. $\{9\}$
    D. $\{0\}$
    E. No real solution

26. Find the distance between the points (2, 5, −2) and (−1, 0, 4).
    A. $\sqrt{30}$
    B. $\sqrt{70}$
    C. 30
    D. 70
    E. 100

27. **Find the equation of a line that contains the point (0,6) and is perpendicular to $2x + y = 4$**
    A. $2x + y = 6$
    B. $x + 2y = 6$
    C. $x - 2y = -12$
    D. $y = -6$
    E. $x = -\dfrac{1}{4}$

28. **Find the equation for the line of symmetry of the parabola $y = 2(x - 3)^2 + 4$.**
    A. $y = x - 3$
    B. $y = 4$
    C. $y = 2$
    D. $x = 2$
    E. $x = 3$

29. **Find the intersection point(s) of** $\begin{cases} x^2 + y^2 = 16 \\ \dfrac{x^2}{16} + \dfrac{y^2}{9} = 1 \end{cases}$
    A. (16, 0)
    B. (0, 16)
    C. (±4, 0)
    D. (0, ±4)
    E. Both A and B

30. **Which quadrant contains the polar coordinate $\left(-2, -\dfrac{\pi}{6}\right)$?**
    A. I
    B. II
    C. III
    D. IV
    E. The point is on the origin.

31. **Given a circle, centered on the origin, with radius 6, which equation below represents moving that circle 3 units to the right and 5 units down?**
    A. $x^2 - y^2 = 36$
    B. $3x^2 - 5y^2 = 36$
    C. $(x - 3)^2 + (y + 5)^2 = 36$
    D. $(x + 3)^2 + (y - 5)^2 = 36$
    E. $\dfrac{x^2}{9} + \dfrac{y^2}{25} = 1$

32. What is the magnitude, in Newtons, of the resulting force on an object when a horizontal force pushes the object with 23N of force while gravity exerts 10N in the vertical direction?
    A. 33
    B. 30.9
    C. 25.1
    D. 23
    E. 13

33. If the surface area of a cube is doubled, what is the resulting change in volume?
    A. The volume is doubled.
    B. The volume is increased by a factor of $2\sqrt{2}$.
    C. The volume is four times greater.
    D. The volume is eight times greater.
    E. The volume stays the same.

34. If a spherical tank has a volume of $288\pi$ cm³, what is the diameter, in cm, of the sphere?
    A. 6
    B. 6.6
    C. 12
    D. 13.2
    E. 24

35. Find the horizontal component of the vector pictured below.

    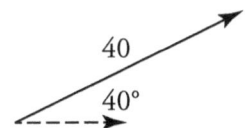

    A. 20
    B. 25.7
    C. 30.6
    D. $40\sqrt{2}$
    E. $40\sqrt{3}$

36. Solve for n in the triangle below.

    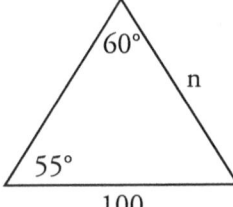

A. 57.7
B. 65
C. 81.9
D. 94.6
E. 105

37. **Solve for θ over [0, 2π]: sin 2θ = sin θ**

    A. $0, \pi, 2, \pi$

    B. $0, \dfrac{\pi}{3}, \pi, \dfrac{5\pi}{3}, 2\pi$

    C. $0, \dfrac{\pi}{2}, \pi, \dfrac{3\pi}{2}, 2\pi$

    D. $\dfrac{\pi}{6}, \dfrac{5\pi}{6}, \dfrac{7\pi}{6}, \dfrac{11\pi}{6}$

    E. No solution

38. **Which equation below represents a periodic graph that contains the origin, has an amplitude of 4 and a period of 6π?**

    A. $y = 4\sin\left(\dfrac{\theta}{3}\right)$

    B. $y = 4\sin(6\theta)$

    C. $y = 4\cos\left(\dfrac{\theta}{3}\right)$

    D. $y = \cos(\theta + 6\pi) + 4$

    E. $y = \sin(3\theta) + 4$

39. **Multiply and simplify:** $(\sin\theta + \cos\theta)^2$

    A. $1$
    B. $\sin^2\theta + \cos^2\theta$
    C. $1 + \tan^2\theta$
    D. $1 + 2\sin\theta$
    E. $1 + \sin 2\theta$

40. **Simplify** $\tan^2\theta - \sin^2\theta$
    A. 1
    B. $\tan^2\theta$
    C. $\sec^2\theta$
    D. $\tan\theta + \sin\theta$
    E. $\tan^2\theta(\sin^2\theta)$

41. **Find the exact value of cos(75°)**
    A. $\dfrac{\sqrt{5}}{2}$
    B. $\dfrac{\sqrt{2}+\sqrt{3}}{2}$
    C. $\dfrac{\sqrt{6}-\sqrt{2}}{4}$
    D. 0.26
    E. 0

42. **Which of the following is undefined?**
    A. $\tan\left(\dfrac{\pi}{2}\right)$
    B. $\csc\left(\dfrac{\pi}{2}\right)$
    C. $\sin^{-1}(-1)$
    D. Both A and B
    E. Both A and C

43. **Convert $\dfrac{7\pi}{5}$ into degrees.**
    A. 4.396°
    B. 75°
    C. 175°
    D. 252°
    E. 285°

44. A group of 5 people in a room represent the following ages: 40, 32, 50, 33, and 43. If a 49 year old enters the room, which of the following will NOT happen?
   A. The mean age will rise.
   B. The median age will rise.
   C. The mode will not change.
   D. There will be an outlier.
   E. There will be an even number of people in the group.

45. Given a jar containing 2 red marbles, 3 white, and 8 black, what is the probability of selecting one white, replacing it, and then reaching in to select one white a second time?
   A. $\dfrac{1}{2}$
   B. $\dfrac{3}{13}$
   C. $\dfrac{6}{13}$
   D. $\dfrac{6}{169}$
   E. $\dfrac{9}{169}$

46. Given a jar containing 2 red marbles, 3 white, and 8 black, what is the probability of selecting a handful of 2 white marbles?
   A. $\dfrac{2}{3}$
   B. $\dfrac{3}{13}$
   C. $\dfrac{9}{169}$
   D. $\dfrac{3}{78}$
   E. $\dfrac{6}{169}$

47. A set of data has a mean of 78 and a standard deviation of 8. Which piece of data from the choices below is within 2 standard deviations of the mean?
   A. 16
   B. 66
   C. 93
   D. Both B and C
   E. All of the above

48. Which equation choice is a reasonable equation for the line of regression through the data points pictured below?

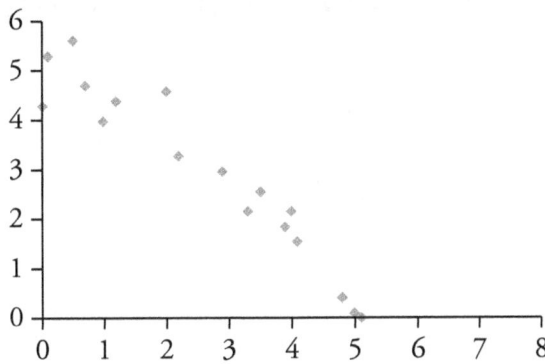

A. $y = 5$
B. $y = x$
C. $y = x + 5$
D. $y = -5x$
E. $y = -x + 5$

49. Which of the following statements is true regarding a quadratic regression equation?
A. The equation can be evaluated to make predictions.
B. The graph of the equation does not necessarily contain all of the data points.
C. The graph is the line of best fit.
D. Both A and B are true.
E. All of the above are true.

50. Given the box and whisker plot below, which of the following statements regarding the plot or its corresponding data is false?

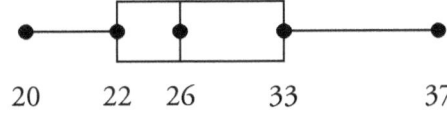

A. The plot represents 5 pieces of data.
B. The median of the data is 26.
C. The largest piece of data is 37.
D. The interquartile range is 11.
E. None of the statements above are false.

# SAT Math 2 Practice Test 2—Answer Key

| | | | | |
|---|---|---|---|---|
| 1. D | 11. B | 21. A | 31. C | 41. C |
| 2. D | 12. D | 22. E | 32. C | 42. A |
| 3. B | 13. C | 23. E | 33. B | 43. D |
| 4. C | 14. A | 24. E | 34. C | 44. D |
| 5. A | 15. B | 25. C | 35. C | 45. E |
| 6. A | 16. E | 26. B | 36. D | 46. D |
| 7. B | 17. E | 27. C | 37. B | 47. D |
| 8. A | 18. B | 28. E | 38. A | 48. E |
| 9. D | 19. E | 29. C | 39. E | 49. D |
| 10. C | 20. C | 30. B | 40. E | 50. A |

# SAT Math Level 2 Practice Test 2
## Answers and Rationales

The solutions presented represent one way to find the answer to the question.

1. Which of the following expressions is not equivalent to $\frac{a+b}{c}$?
   A. $\frac{1}{c}(a+b)$
   B. $\frac{a}{c}+\frac{b}{c}$
   C. $\frac{b+a}{c}$
   D. $a+b \div c$
   E. All of the above are equivalent to the given expression.

Answer: D

In order for D to be an equivalent expression, $a + b$ would need to be in parenthesis

2. The dosage of a certain antibiotic must be measured as 40 mg of medicine for every 25 lb of patient weight. How many milligrams must be prescribed for a 140-lb patient?
   A. 5 mg
   B. 165 mg
   C. 180 mg
   D. 224 mg
   E. 250 mg

Answer: D

Set up and solve a proportion: $\frac{\text{mg of medicine}}{\text{patient weight in lb}} \rightarrow \frac{40}{25} = \frac{x}{140}$

$$25x = 40(140)$$
$$x = 224$$

3. How many different committees can be formed by selecting 4 members from a pool of 50 candidates?
   A. 5,527,200
   B. 230,300
   C. 200
   D. 54
   E. 4

**Answer: B**

The combination $_{50}C_4$ represents this group of 4 people whose order does not matter. A calculator can be used to find this value; otherwise use the corresponding factorial
$$\frac{50!}{4!(50-4)!}$$

4. Find $\sum_{n=1}^{5} n^2$
   A. 24
   B. 25
   C. 55
   D. 100
   E. None of the above

**Answer: C**

The given notation represents the sum of the squares of the first 5 counting numbers.
$$1^2 + 2^2 + 3^2 + 4^2 + 5^2 = 55$$

5. **If n represents any whole number, which expression below represents the product of 2 consecutive, odd numbers?**
   A. $(2n + 1)(2n+3)$
   B. $(n + 1)(n + 3)$
   C. $(3n)(5n)$
   D. $n(n^2)$
   E. $n(n+2)$

**Answer: A**

Since one greater than an even number is always odd, and since $2n$ will always be even, $2n + 1$ will always be odd. The next, consecutive odd is 2 greater so $2n + 1 + 2 = 2n + 3$ represents the next odd.

6. **Simplify $2^3 i^5$**
   A. $8i$
   B. $-8$
   C. $6i$
   D. $-6i$
   E. $\sqrt{-8}$

**Answer: A**

The real component of the expression, $2^3$, is 8. The imaginary portion should be broken down as follows:
$i^5 = i^4 \cdot i = (1)i = i$

7. **Find the magnitude of the vector 6i + 8j.**
   A. $8\sqrt{3}$
   B. 10
   C. 14
   D. 48

E. None of the above

**Answer: B**

The vector is given in terms of the standard unit vectors *i* and *j*, with a horizontal component of 6 and a vertical component of 8. Double the Pythagorean triple, 3, 4, 5, to find the hypotenuse, or vector magnitude, to be 10.

8. **Write an expression representing the following relationship: "double the square of a number."**
   A. $2n^2$
   B. $(2n)^2$
   C. $2(25n)$
   D. $n^4$
   E. Both A and B

**Answer: A**

The square of the number is represented by $n^2$, and multiplication by 2 doubles that square.

9. **Simplify the expression $(4x^8y^5)(4xy^3)^{-2}$**

   A. $x^7y^2$

   B. $-16x^6y$

   C. $\dfrac{x^4}{12y}$

   D. $\dfrac{x^6}{4y}$

   E. $\dfrac{x^6y}{4}$

**Answer: D**

First rewrite the expression with a positive exponent:

$\dfrac{4x^8y^5}{(4xy^3)^2} = \dfrac{4x^8y^5}{16x^2y^6} = \dfrac{x^6}{4y}$ (Subtract exponents on like bases and reduce the fraction $\dfrac{4}{16}$)

10. **Solve for x:** $\dfrac{1}{3}x + 2 = \dfrac{3}{5}x + \dfrac{1}{3}$

    A. $\dfrac{2}{3}$

    B. $\dfrac{4}{5}$

    C. $\dfrac{25}{4}$

    D. $\dfrac{35}{3}$

E. $\dfrac{48}{5}$

**Answer: C**

Multiply both sides of the equation by 15, the least common denominator, to clear the fractions.
$$5x + 30 = 9x + 5$$
$$25 = 4x$$
$$x = \dfrac{25}{4}$$

11. **Solve over the complex numbers:** $9x^2 + 49 = 0$

    A. $\pm 2\sqrt{10}$

    B. $\pm \dfrac{7i}{3}$

    C. $\dfrac{7}{3}$

    D. $\dfrac{49}{9}$

    E. $-40$

**Answer: B**

First solve for $x^2$, then take the plus or minus square root of both sides.
$$9x^2 = -49$$
$$x^2 = \dfrac{-49}{9}$$
$$x = \pm\sqrt{\dfrac{-49}{9}} = \dfrac{7i}{3}$$

The imaginary number, $i$, is the complex result of taking the square root of a negative number.

12. **Which system of equations below has an infinite number of solutions?**

    A. $\begin{cases} 5x + y = 8 \\ 3x - 4y = 14 \end{cases}$

    B. $\begin{cases} 2x + y = 7 \\ y = 4 \end{cases}$

    C. $\begin{cases} 3x - 2y = 8 \\ 6x - 4y = 8 \end{cases}$

    D. $\begin{cases} x + y = 12 \\ 5x + 5y = 60 \end{cases}$

    E. $\begin{cases} x^2 + y^2 = 25 \\ x + y = 5 \end{cases}$

**Answer: D**

The two equations in choice D represent the same line, as the second is the first multiplied by 5. The infinite solutions, then, are all the points along the line $x + y = 12$.

13. **Solve for x: x3 - 3x2 - 3x + 9 = 0**
    A. 0, 3
    B. −2, 2, 3
    C. $-\sqrt{3}, \sqrt{3}, 3$
    D. −3, 3, 9
    E. 0, 3, 9

**Answer: C**

One method of solution is to graph the polynomial to find one real root, in this case 3, and then use synthetic division to find the remaining two solutions. This particular polynomial, however, can be factored to reveal the solutions.

$$x^3 - 3x^2 - 3x + 9 = 0$$
$$x^2(x-3) - 3(x-3) = 0$$
$$(x-3)(x^2 - 3) = 0$$
$$x - 3 = 0, \; x^2 - 3 = 0$$
$$x = 3, \quad x^2 = 3$$
$$x = \pm\sqrt{3}$$

14. **Solve for x. Round the answer to the nearest hundredth. 3x = 12**
    A. 2.26
    B. 3.14
    C. 4.00
    D. 4.12
    E. 6.00

**Answer: A**

Rewrite the exponential equation in logarithmic form: $\log_3 12 = x$.
Then estimate the log value directly on a calculator, or use the change of base rule for a calculator that only gives values of log with base 10: $\log_3 12 = \dfrac{\log 12}{\log 3} \approx 2.26$

15. **Solve the compound inequality: $7 \leq 3x + 1 \leq 49$**
    A. $0 \leq x \leq 2$
    B. $2 \leq x \leq 16$
    C. $3.5 \leq x \leq 24.5$
    D. $x \leq 2$ or $x \leq 16$
    E. $x \geq 0$ or $x \leq 2$

**Answer: B**

Solve the inequality by performing the same operation to all three sections.
$$7 \leq 3x + 1 \leq 49$$
$$6 \leq 3x \leq 48$$
$$2 \leq x \leq 16$$

16. **Solve for x, such that $0° \leq x \leq 360°$: 8 sin x + 1 = 5**

    A. $\frac{1}{2}$

    B. 1.73

    C. 30

    D. 150

    E. Both C and D

**Answer: E**

First isolate the trigonometric function:
$$8 \sin x = 4$$
$$\sin x = \frac{4}{8} = \frac{1}{2}$$

Then determine which angle values have a sine of $\frac{1}{2}$. This occurs two places on the unit circle: in the first quadrant, at 30°, and in the second quadrant at 150°.

17. **Which function listed below is not defined for x = 0, −2?**

    A. $f(x) = x^2 + 2x$

    B. $g(x) = \sqrt{x^2 + 2x}$

    C. $p(x) = x(x - 2)$

    D. $q(x) = 5$

    E. $h(x) = \dfrac{3}{x^2 + 2x}$

**Answer: E**

When evaluating $h(x)$ for $x = 0$ or $-2$, the denominator will become zero which makes the function undefined.

18. **If $h(x) = x^2$ and $g(x) = x + 3$, which statement below is false?**

    A. $g(h(x)) = x^2 + 3$

    B. $h(g(x)) = x^2 + 9$

    C. $(h \circ g)(x) = x^2 + 6x + 9$

    D. $h(x) \bullet g(x) = x^3 + 3x^2$

    E. B and D are both false

**Answer: B**

The notation $h(g(x))$ signifies the same function composition as $(h \circ g)(x)$. To properly calculate the function composition: $(h \circ g)(x) = (x+3)^2 = x^2 + 6x + 9$ when the binomial is squared correctly. Additionally, choice D correctly represents the multiplication of the two functions.

19. **Which equation below represents a function with zeros −1, 2, and 7?**
    A. $f(x) = x^7 + x^2 - x$
    B. $g(x) = x^2 + 7x - 1$
    C. $h(x) = 2x + 7$
    D. $m(x) = 2x^3 - x^2 + 7x$
    E. $p(x) = x^3 - 8x^2 + 5x + 14$

**Answer: E**

If the zeros of the function are −1, 2, and 7, then the factors are $(x + 1)(x - 2)(x - 7)$. Multiplying the three factors together yields choice E.

20. **Find the Cartesian equation that corresponds to the given set of parametric equations.**

    $$\begin{cases} x(t) = 5t \\ y(t) = t^2 + 1 \end{cases}$$

    A. $y = x^2 + 5x + 1$

    B. $y = 5x^2 + 5$

    C. $y = \dfrac{x^2}{25} + 1$

    D. $y = 25x^2 + 1$

    E. $y = \dfrac{1}{5}x$

**Answer: C**

Solve one of the parametric equations for $t$, and use it to substitute into the other.

$$x = 5t \rightarrow t = \frac{x}{5}$$

$$y = t^2 + 1 = \left(\frac{x}{5}\right)^2 + 1$$

$$y = \frac{x^2}{25} + 1$$

21. **If** $f(x) = \begin{cases} 2x \text{ for } x < 0 \\ \sqrt{x} \text{ for } x \geq 0 \end{cases}$, **find f(−16)**

    A. −32
    B. −16
    C. 0

D. 4

E. 4*i*

**Answer: A**

Since the input for the function is less than zero, use the top portion of the piecewise rule.
$f(x) = 2x = 2(-16) = -32$

22. **If h(x) and g(x) are inverses of each other, then which statement below is true?**

   A. $h(x) = -g(x)$

   B. $h(x) = \dfrac{1}{g(x)}$

   C. $h(g(a)) = a$

   D. $g(h(a)) = a$

   E. Both C and D are true

**Answer: E**

If a function is applied to an input and then the inverse function is applied, the output reverts to the original input.

23. **Simplify** $\left(64x^{10}y^{-2}\right)^{3/2}$

   A. $\sqrt{64x^{10}y^2}$

   B. $8x^5 y\sqrt{3}$

   C. $96x^{15}y^3$

   D. $-512x^{15}y^3$

   E. $\dfrac{512x^{15}}{y^3}$

**Answer: E**

Apply the exponent to each factor of the contents of the parenthesis, and multiply exponents.

$$\left(8^2\right)^{3/2} \left(x^{10}\right)^{3/2} \left(y^{-2}\right)^{3/2}$$
$$8^3 x^{15} y^{-3}$$
$$\dfrac{8^3 x^{15}}{y^3}$$

24. **Solve for x: $x^2 + 4x + 5 = 0$**

    A. {4, 5}

    B. {1, 5}

    C. $\{2, \sqrt{5}\}$

    D. $\dfrac{4 \pm \sqrt{26}}{2}$

    E. No real solution

**Answer: E**

As the trinomial is not factorable, the quadratic formula can be used to find a solution. The value of the discriminant, however, $b^2 - 4ac = 4^2 - 4(1)(5) = -4$ is negative, which indicates that there is no real solution to this equation.

25. **Solve $x - 5 = \sqrt{x + 7}$**

    A. {5, 7}

    B. {2, 9}

    C. {9}

    D. {0}

    E. No real solution

**Answer: C**

To solve a radical equation, raise each side to the inverse power, in this case, the power of 2.

$$(x-5)^2 = (\sqrt{x+7})^2$$
$$x^2 - 10x + 25 = x + 7$$
$$x^2 - 11x + 18 = 0$$
$$(x-9)(x-2)$$
$$x = 9, 2$$

However, when solving a radical equation, extraneous solutions can arise so the solutions must be checked in the original equation. The value −2 does not check.

$$2 - 5 \neq \sqrt{2 + 7}$$
$$-3 \neq \sqrt{9}$$

The 9 does check and remains the lone solution to the problem.

26. **Find the distance between the points (2, 5, –2) and (–1, 0, 4).**
    A. $\sqrt{30}$
    B. $\sqrt{70}$
    C. 30
    D. 70
    E. 100

Answer: B

The distance formula for points in three dimensional space is comparable to that in two dimensions. Simply find the square root of the sum of the squares of the differences between each coordinate.
$$\sqrt{(2-(-1))^2 + (5-0)^2 + (-2-4)^2}$$
$$\sqrt{3^2 + 5^2 + (-6)^2}$$
$$\sqrt{9 + 25 + 36} = \sqrt{70}$$

27. **Find the equation of a line that contains the point (0,6) and is perpendicular to $2x + y = 4$**
    A. $2x + y = 6$
    B. $x + 2y = 6$
    C. $x – 2y = –12$
    D. $y = –6$
    E. $x = -\dfrac{1}{4}$

Answer: C

Solve the given line for y to find its slope: $y = –2x + 4$. The slope is –2. Then any line perpendicular to this must have a slope of ½. With a requested y intercept of 6 and slope of ½, the new line's equation starts as $y = \dfrac{1}{2}x + 6$ which, when put in standard form, is represented by choice C.

28. **Find the equation for the line of symmetry of the parabola $y = 2(x - 3)^2 + 4$.**
    A. $y = x – 3$
    B. $y = 4$
    C. $y = 2$
    D. $x = 2$
    E. $x = 3$

Answer: E

The given equation represents a parabola, opening up, with vertex (3, 4). A vertical line of symmetry goes through the vertex making choice E the correct answer.

29. **Find the intersection point(s) of** $\begin{cases} x^2 + y^2 = 16 \\ \dfrac{x^2}{16} + \dfrac{y^2}{9} = 1 \end{cases}$
    A. (16, 0)
    B. (0, 16)
    C. (±4, 0)
    D. (0, ±4)

E. Both A and B

**Answer: C**

To find the solution algebraically, solve the first equation for $y^2$, and substitute into the second equation.

$$y^2 = 16 - x^2$$
$$\frac{x^2}{16} + \frac{16-x^2}{9} = 1$$
$$9x^2 + 16(16 - x^2) = 144$$
$$-7x^2 = -112$$
$$x^2 = 16$$
$$x = \pm 4$$

Confirm this solution graphically by knowing the first equation is a circle, centered on the origin, with radius 4 and the second equation represents an ellipse, also centered on the origin, with a horizontal major axis of length 8. The diameter of the circle, then, is the same as the major axis of the ellipse so the shapes intersect at each end: (4, 0) and (−4, 0).

30. **Which quadrant contains the polar coordinate $\left(-2, -\frac{\pi}{6}\right)$?**
    A. I
    B. II
    C. III
    D. IV
    E. The point is on the origin.

**Answer: B**

Polar coordinates are given in the form (r, θ) where r is a radial distance from the origin and θ is an angle of rotation. The angle $-\frac{\pi}{6}$ equates to −30°, which is in Quadrant IV. However, the given, negative radius directs the point "backwards" and it ends up diametrically opposite, in Quadrant II.

31. **Given a circle, centered on the origin, with radius 6, which equation below represents moving that circle 3 units to the right and 5 units down?**
    A. $x^2 - y^2 = 36$
    B. $3x^2 - 5y^2 = 36$
    C. $(x - 3)^2 + (y + 5)^2 = 36$
    D. $(x + 3)^2 + (y - 5)^2 = 36$
    E. $\frac{x^2}{9} + \frac{y^2}{25} = 1$

**Answer: C**

Moving the circle as directed gives the circle a center at (3, −5). The radius remains 6. The standard form for a circle with radius $r$ and center $(h, k)$ is $(x - h)^2 + (y - k)^2 = r^2$. Therefore, choice C is the correct answer.

32. **What is the magnitude, in Newtons, of the resulting force on an object when a horizontal force pushes the object with 23N of force while gravity exerts 10N in the vertical direction?**
   A. 33
   B. 30.9
   C. 25.1
   D. 23
   E. 13

Answer: C

Create a head-to-tail vector diagram for the given forces and use the Pythagorean Theorem to solve for the magnitude of the resulting vector. $23^2 + 10^2 = m^2$, $m \approx 25.1$

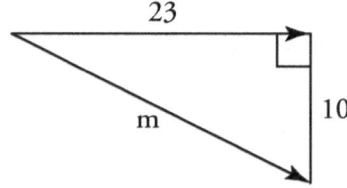

33. **If the surface area of a cube is doubled, what is the resulting change in volume?**
   A. The volume is doubled.
   B. The volume is increased by a factor of $2\sqrt{2}$.
   C. The volume is four times greater.
   D. The volume is eight times greater.
   E. The volume stays the same.

Answer: B

Start with a cube of side $n$, whose surface area, then, is $6n^2$ and volume is $n^3$. Next, double that surface area: $2(6n^2) = 12n^2$. Find the length, $s$, of a side of this new cube in terms of the old length, $n$:

$$6s^2 = 12n^2$$
$$s^2 = 2n^2$$
$$s = n\sqrt{2}$$

Then use the new length to calculate the new volume: $V = s^3 = \left(n\sqrt{2}\right)^3 = 2n^3\sqrt{2}$

34. **If a spherical tank has a volume of $288\pi$ cm³, what is the diameter, in cm, of the sphere?**
   A. 6
   B. 6.6
   C. 12
   D. 13.2
   E. 24

**Answer: C**

The formula for the volume of a sphere with radius $r$ is $\frac{4}{3}\pi r^3$. Solve the equation for $r$.

$$V = \frac{4}{3}\pi r^3$$
$$288\pi = \frac{4}{3}\pi r^3$$
$$288 = \frac{4}{3}r^3$$
$$216 = r^3$$
$$r = 6$$

This means the radius is 6. Double to find the diameter of 12.

**35. Find the horizontal component of the vector pictured below.**

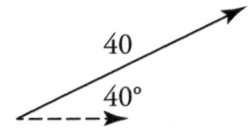

A. 20
B. 25.7
C. 30.6
D. $40\sqrt{2}$
E. $40\sqrt{3}$

**Answer: C**

Use a triangle diagram of the vector to set up a trigonometry equation to find the horizontal component, $x$.

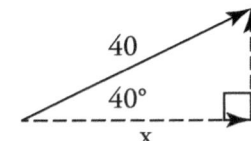

$$\cos(40) = \frac{x}{40}$$
$$40\cos(40) = x$$
$$x \approx 30.6$$

**36.** Solve for n in the triangle below.

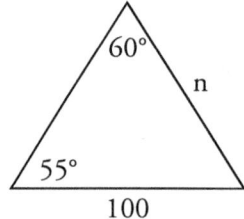

A. 57.7
B. 65
C. 81.9
D. 94.6
E. 105

**Answer: D**

Use the Law of Sines to set up a proportion.

$$\frac{\sin(60)}{100} = \frac{\sin(55)}{n}$$
$$n\sin(60) = 100\sin(55)$$
$$n = \frac{100\sin(55)}{\sin(60)} \approx 94.6$$

**37.** Solve for θ over [0, 2π]: sin 2θ = sin θ

A. 0, π, 2, π

B. $0, \frac{\pi}{3}, \pi, \frac{5\pi}{3}, 2\pi$

C. $0, \frac{\pi}{2}, \pi, \frac{3\pi}{2}, 2\pi$

D. $\frac{\pi}{6}, \frac{5\pi}{6}, \frac{7\pi}{6}, \frac{11\pi}{6}$

E. No solution

**Answer: B**

Set the equation equal to zero, then use the double angle identity.

$$\sin 2\theta - \sin \theta = 0$$
$$2\sin\theta\cos\theta - \sin\theta = 0$$
$$\sin\theta(2\cos\theta - 1) = 0$$
$$\sin\theta = 0 \quad \text{or} \quad 2\cos\theta - 1 = 0$$
$$\theta = 0, \pi, 2\pi \qquad \cos\theta = \frac{1}{2}$$
$$\theta = \frac{\pi}{3}, \frac{5\pi}{3}$$

38. **Which equation below represents a periodic graph that contains the origin, has an amplitude of 4 and a period of 6π?**

   A. $y = 4\sin\left(\dfrac{\theta}{3}\right)$

   B. $y = 4\sin(6\theta)$

   C. $y = 4\cos\left(\dfrac{\theta}{3}\right)$

   D. $y = \cos(\theta + 6\pi) + 4$

   E. $y = \sin(3\theta) + 4$

**Answer: A**

The basic graph of $y = \sin\theta$ contains the origin, has a period of $2\pi$, and an amplitude of 1. The external factor of 4 increases the amplitude directly, and the internal factor changes the period inversely. For instance, in choice A, the internal factor is $\frac{1}{3}$, so the period is $\frac{2\pi}{1/3} = 6\pi$.

39. **Multiply and simplify:** $(\sin\theta + \cos\theta)^2$
   A. 1
   B. $\sin^2\theta + \cos^2\theta$
   C. $1 + \tan^2\theta$
   D. $1 + 2\sin\theta$
   E. $1 + \sin 2\theta$

**Answer: E**
First square the binomial. Then apply the Pythagorean and double angle identities.
$$\sin^2\theta + 2\sin\theta\cos\theta + \cos^2\theta$$
$$\sin^2\theta + \cos^2\theta + 2\sin\theta\cos\theta$$
$$1 + 2\sin\theta\cos\theta$$
$$1 + \sin 2\theta$$

**40. Simplify** $\tan^2\theta - \sin^2\theta$
   A. 1
   B. $\tan^2\theta$
   C. $\sec^2\theta$
   D. $\tan\theta + \sin\theta$
   E. $\tan^2\theta(\sin^2\theta)$

**Answer: E**
First rewrite tangent in terms of sine and cosine. Then combine fractions with like denominators.
$$\frac{\sin^2\theta}{\cos^2\theta} - \sin^2\theta$$
$$\frac{\sin^2\theta}{\cos^2\theta} - \frac{\sin^2\theta(\cos^2\theta)}{\cos^2\theta}$$
$$\frac{\sin^2\theta(1 - \cos^2\theta)}{\cos^2\theta}$$
$$\frac{\sin^2\theta(\sin^2\theta)}{\cos^2\theta}$$
$$\sin^2\theta\left(\frac{\sin^2\theta}{\cos^2\theta}\right)$$
$$\sin^2\theta(\tan^2\theta)$$

**41. Find the exact value of** $\cos(75°)$
   A. $\dfrac{\sqrt{5}}{2}$
   B. $\dfrac{\sqrt{2}+\sqrt{3}}{2}$
   C. $\dfrac{\sqrt{6}-\sqrt{2}}{4}$
   D. 0.26
   E. 0

**Answer: C**

While choice D is an approximation of the cosine value to the nearest hundredth, the exact value is found by applying the trigonometric sum identity.

$$\cos(75) = \cos(45+30)$$
$$\cos(45)\cos(30) - \sin(45)\sin(30)$$
$$\frac{\sqrt{2}}{2}\left(\frac{\sqrt{3}}{2}\right) - \frac{\sqrt{2}}{2}\left(\frac{1}{2}\right)$$
$$\frac{\sqrt{6}}{4} - \frac{\sqrt{2}}{4}$$

42. **Which of the following is undefined?**

   A. $\tan\left(\dfrac{\pi}{2}\right)$

   B. $\csc\left(\dfrac{\pi}{2}\right)$

   C. $\sin^{-1}(-1)$

   D. Both A and B

   E. Both A and C

**Answer: A**

Since $\tan\left(\dfrac{\pi}{2}\right) = \dfrac{\sin\left(\dfrac{\pi}{2}\right)}{\cos\left(\dfrac{\pi}{2}\right)}$, and $\cos\left(\dfrac{\pi}{2}\right) = 0$, the value for A is undefined. Choices B and C are both reasonable values answering as 1 and $\dfrac{3\pi}{2}$ respectively.

43. **Convert $\dfrac{7\pi}{5}$ into degrees.**

   A. 4.396°
   B. 75°
   C. 175°
   D. 252°
   E. 285°

**Answer: D**

Convert $\dfrac{7\pi}{5} \text{rad}\left(\dfrac{180 \text{ deg}}{\pi \text{ rad}}\right) = \dfrac{7}{5}(180 \text{ deg}) = 252°$

44. **A group of 5 people in a room represent the following ages: 40, 32, 50, 33, and 43. If a 49 year old enters the room, which of the following will NOT happen?**
   A. The mean age will rise.
   B. The median age will rise.
   C. The mode will not change.
   D. There will be an outlier.
   E. There will be an even number of people in the group.

Answer: D

Since the number 49 is greater than both the mean and median of the original set of data, both A and B are true. And since no values are repeated either before or after the 49 year old enters, choice C is true. Assuming there is data to represent each group member, choice E is correct. Choice D is not true, because the value 49 is reasonably close to the existing data. No extremes or outliers exist.

45. **Given a jar containing 2 red marbles, 3 white, and 8 black, what is the probability of selecting one white, replacing it, and then reaching in to select one white a second time?**
   A. $\frac{1}{2}$
   B. $\frac{3}{13}$
   C. $\frac{6}{13}$
   D. $\frac{6}{169}$
   E. $\frac{9}{169}$

Answer: E

The probability of selecting one white marble is 3 out of 13, or $\frac{3}{13}$. To repeat this as an independent event is to multiply the probabilities. $\frac{3}{13} \cdot \frac{3}{13} = \frac{9}{169}$

46. **Given a jar containing 2 red marbles, 3 white, and 8 black, what is the probability of selecting a handful of 2 white marbles?**
   A. $\frac{2}{3}$
   B. $\frac{3}{13}$
   C. $\frac{9}{169}$
   D. $\frac{3}{78}$
   E. $\frac{6}{169}$

**Answer: D**

To grab a handful is to calculate combinations. How many ways can 2 whites out of the 3 be chosen? How many ways can a handful of 2 marbles be chosen out of the total 13?

$$\frac{_3C_2}{_{13}C_2} = \frac{3}{78}$$

47. A set of data has a mean of 78 and a standard deviation of 8. Which piece of data from the choices below is within 2 standard deviations of the mean?
    A. 16
    B. 66
    C. 93
    D. Both B and C
    E. All of the above

**Answer: D**

A standard deviation, σ, away from the mean can go in the positive as well as the negative direction. As illustrated in the number line, both 66 and 93 are within 2 standard deviation lengths of the mean.

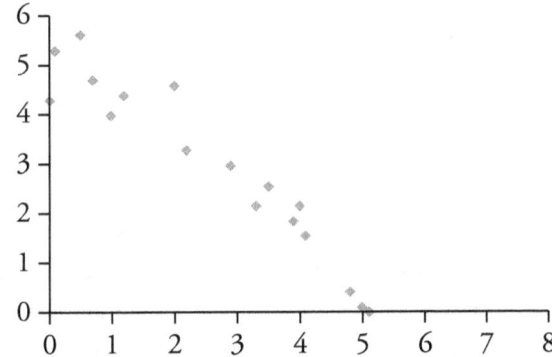

48. Which equation choice is a reasonable equation for the line of regression through the data points pictured below?

    A. $y = 5$
    B. $y = x$
    C. $y = x + 5$
    D. $y = -5x$
    E. $y = -x + 5$

**Answer: E**

The best fit line drawn through the data points will have a negative slope and have intercepts close to (0, 5) and (5, 0). This makes choice E the best approximation for the regression line out of the choices given, as it contains the point (0, 5) and has a slope of −1.

49. **Which of the following statements is true regarding a quadratic regression equation?**
    A. The equation can be evaluated to make predictions.
    B. The graph of the equation does not necessarily contain all of the data points.
    C. The graph is the line of best fit.
    D. Both A and B are true.
    E. All of the above are true.

**Answer: D**

Choice C is not true for a quadratic regression equation as the line of best fit refers to a *linear* regression equation. A quadratic regression equation can be described as the best fit *curve* through the points. It is drawn through most of the points but does not always contain every data point. Once the equation is determined, it can be used to make predictions as to where new data points might fall.

50. **Given the box and whisker plot below, which of the following statements regarding the plot or its corresponding data is false?**

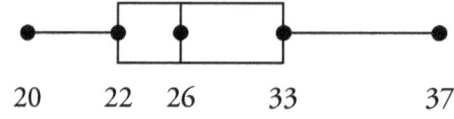

    A. The plot represents 5 pieces of data.
    B. The median of the data is 26.
    C. The largest piece of data is 37.
    D. The interquartile range is 11.
    E. None of the statements above are false.

**Answer: A**

The box and whisker plot is built as a number line highlighting the minimum and maximum of the data (here seen as 20 and 37, respectively), as well as the three quartile values (in this case 22, 26, and 33). An indeterminate number of data pieces exist to create this spread that are not shown on the box and whisker plot. We only know that half of the data, for instance, lies between quartile 1 and 3, but we do not know how many pieces of data there are.

# SAT

SAT Subject Tests are college admission exams on specific subjects. These tests are generally given six times in any given school year, on the same days and in the same test centers as the SAT — but not all 20 tests are offered on every SAT date. When you take an SAT Subject Test, you are doing more than showing off your strengths.

If the college decides to give you credit, it will record the number of credits on your permanent record, thereby indicating that you have completed work equivalent to a course in that subject. If the college decides to grant exemption without giving you credit for a course, you will be permitted to omit a course that would normally be required of you and to take a course of your choice instead.

**SAT Math 1**
ISBN 978-1-60787-571-0
$16.99

**SAT Math 2**
ISBN 978-1-60787-572-7
$14.99

**SAT Biology**
ISBN 978-1-60787-569-7
$18.99

**SAT Chemistry**
ISBN 978-1-60787-568-0
$14.99

**SAT Literature**
ISBN 978-1-60787-573-4
$16.99

**SAT Spanish**
ISBN 978-1-60787-570-3
$19.99

or amazon or BARNES & NOBLE BOOKSELLERS

XAMonline.com

www.ingramcontent.com/pod-product-compliance
Lightning Source LLC
LaVergne TN
LVHW061308060426
835507LV00019B/2066